ST. MARY'S COLLEGE OF MARYLAND LIBRARY
ST. MARY'S CITY, MARYLAND

C65076

Probability and Mathematical Statistics

AN INTRODUCTION

PROBABILITY AND MATHEMATICAL STATISTICS

An Introduction

EUGENE LUKACS
The Catholic University of America

ACADEMIC PRESS   New York and London

COPYRIGHT © 1972, BY ACADEMIC PRESS, INC.
ALL RIGHTS RESERVED
NO PART OF THIS BOOK MAY BE REPRODUCED IN ANY FORM,
BY PHOTOSTAT, MICROFILM, RETRIEVAL SYSTEM, OR ANY
OTHER MEANS, WITHOUT WRITTEN PERMISSION FROM
THE PUBLISHERS.

ACADEMIC PRESS, INC.
111 Fifth Avenue, New York, New York 10003

United Kingdom Edition published by
ACADEMIC PRESS, INC. (LONDON) LTD.
24/28 Oval Road, London NW1 7DD

LIBRARY OF CONGRESS CATALOG CARD NUMBER: 74-154395

AMS(MOS) 1970 Subject Classifications: 60-01, 62-01

PRINTED IN THE UNITED STATES OF AMERICA

Contents

Preface ix

Introduction 1

Part I. **PROBABILITY THEORY**

Chapter 1. **The Probability Space**

1.1	The Outcome Space	7
1.2	Probabilities	11
1.3	The Axioms	14
1.4	Problems	16
	References	18

Chapter 2. Elementary Properties of Probability Spaces

2.1	Simple Consequences of the Axioms	19
2.2	Conditional Probability and Independence	24
2.3	Finite Probability Spaces	31
2.4	Problems	34

Chapter 3. Random Variables and Their Probability Distributions

3.1	Random Variables	37
3.2	Distribution Functions	40
3.3	Examples of Discrete Distributions	44
3.4	Examples of Absolutely Continuous Distributions	48
3.5	Multivariate Distributions	54
3.6	Problems	67
	References	69

Chapter 4. Typical Values

4.1	The Mathematical Expectation of a Random Variable	71
4.2	Expectations of Functions of Random Variables	76
4.3	Properties of Expectations	81
4.4	Moments	84
4.5	Regression	96
4.6	Problems	100
	Reference	103

Chapter 5. Limit Theorems

5.1	Laws of Large Numbers	105
5.2	The Central Limit Theorem	108
5.3	The Poisson Approximation to the Binomial	113
5.4	Problems	115
	References	117

Chapter 6. Some Important Distributions

6.1	The Distribution of the Sum of Independent, Absolutely Continuous Random Variables	119

6.2 Addition of Independent Normal Random Variables 121
6.3 The Chi-Square Distribution 123
6.4 Student's Distribution 129
6.5 Problems 132

Part II. MATHEMATICAL STATISTICS

Chapter 7. Sampling

7.1 Statistical Data 137
7.2 Sample Characteristics 139
7.3 Moments and Distributions of Sample Characteristics 142
7.4 Problems 149
     References 151

Chapter 8. Estimation

8.1 Properties of Estimates 154
8.2 Point Estimation 157
8.3 Interval Estimation 162
8.4 Problems 169
     References 173

Chapter 9. Testing Hypotheses

9.1 Statistical Hypotheses 176
9.2 The Power of a Test 178
9.3 The t-Test 184
9.4 Nonparametric Methods 186
9.5 Problems 191
     References 197

Appendix A. **Some Combinatorial Formulas** 199

Appendix B. **The Gamma Function** 205

Appendix C.	**Proof of the Central Limit Theorem**	207
Appendix D.	**Tables**	219
Answers to Selected Problems		227
Index		237

Preface

This book is designed for a brief undergraduate course in probability theory and mathematical statistics. The selection of the topics was considerably influenced by the outline of the 2P course suggested by the Committee on the Undergraduate Program of the Mathematical Association of America. The book can, therefore, be used in a 2P course offered to students having only the minimal mathematical prerequisites.

The mathematical prerequisites of this book are modest. Some familiarity with the elements of differential and integral calculus is assumed. However, many topics traditionally taught in the second or third part of a calculus course are deliberately avoided. This fact imposes very serious restrictions on the techniques employed in proving the theorems presented in the book. No use is made of Jacobians, moment generating and characteristic functions, or of matrices and n-dimensional geometry. Students unfamiliar with these topics will not be handicapped in using this book. Partial differentiation is used very sparingly; it occurs only in connection with the discussion of least square regression.

As a consequence of these restrictions it is necessary to omit some proofs and to refer for these to more advanced books. In other cases proofs are adapted to the modest mathematical prerequisites that are available. For example, the chi-square distribution as the distribution of the sum of squares of independent standardized normal variables is derived by mathematical induction. Whereas the central limit theorem for Bernoulli trials is proven in the text, the central limit theorem for independently and identically distributed random variables having finite variances is only stated in the main body of the book. However, Appendix C contains the proof of a more general form of the central limit theorem. This proof does not use characteristic functions but the method of probability operators. The technique of probability operators is suitable for courses at this level. Nevertheless the proof was relegated to the appendix since it is realized that the time allotted for a course using this book will not always allow its presentation.

In view of the modest mathematical prerequisites the book cannot aim at giving a complete and rigorous development of the subject. Yet I do not wish to offer simply results and recipes. My aim is to provide a solid and well-balanced first introduction to probability and mathematical statistics.

I wish to express my thanks to my colleague, Professor E. Batschelet, for reading the manuscript and to Mrs. A. Miller and Mrs. P. Spathelf for typing the manuscript accurately. Thanks are also due to Mr. G. L. Grunkemeier and Mr. M. S. Scott, Jr. for their assistance in reading the proofs.

Probability and Mathematical Statistics

AN INTRODUCTION

Introduction

Certain games of chance, like tossing dice, date back to antiquity but became very popular during the Renaissance. Some gamblers of this period made observations that they could not explain, so they approached several famous scientists† and stimulated them to study games of chance. These scientists' investigations used mathematical tools, but those tools did not yet constitute a mathematical theory. For a long time it remained even doubtful whether probability theory (this was the name given to these studies) would become a part of physics or of mathematics. This question, explicitly formulated by D. Hilbert, was answered early in the second quarter of the twentieth century when A. N. Kolmogorov established a rigorous mathematical foundation for

---

† See the references at the end of this introduction.

probability theory. Since we shall use this approach, it is appropriate to make first a few remarks concerning the structure of a mathematical theory.

A mathematical theory deals with relations among concepts. Certain undefined notions (fundamental concepts) are introduced, and their properties are described by certain propositions that are stated without proof. These propositions are used as fundamental assumptions for the theory. These statements are called *axioms*. A deductive system is derived from the axioms by drawing conclusions either from the axioms or from propositions derived from the axioms. New concepts are introduced by giving precise and exhaustive definitions. The properties of the new concepts are also studied. Every proposition, except the axioms, is established in a strictly deductive manner. The axioms contain therefore all the properties of the fundamental concepts that can be used in constructing the theory. In this sense the ensemble of axioms defines the set of fundamental concepts. This kind of definition is seemingly different from the explicit way in which new concepts are introduced in the later parts of a mathematical theory. However, from a logical viewpoint the axioms serve the same purpose as the definitions: They explain the meaning of the fundamental concepts. The formal difference is that the meaning of all fundamental concepts is given simultaneously by the set of axioms while later each concept is defined explicitly. It is customary to talk about definition by axioms or implicit definitions in distinction to explicit definition. The axioms must of course be consistent, this means that one must be assured that neither the axioms nor their consequences contain a contradiction.

From a strictly logical viewpoint it is possible to choose the axioms arbitrarily. This is however rarely done since an arbitrary selection of axioms would lead to a structure resembling a party game and would lack interest. The selection of axioms is frequently motivated by experience and the wish to build a mathematical model for certain physical phenomena. The system of axioms provides then an idealized description of these phenomena. The fundamental concepts correspond to physical entities, the axioms to their properties. At least some of the derived propositions should correspond to observable facts; of course it may

INTRODUCTION

happen that some consequences cannot be verified—at least with the available technical means.

The study of probability theory was stimulated by observing a peculiar property of games of chance: Their irregularity makes it impossible to predict the outcome of one particular game, yet if one plays a large number of games the situation changes. One can predict average gains or one can state which of two bets is more favorable. The property is shared by other phenomena; as examples we mention the yield of a plot on which a certain crop was planted, the number of defective items in a mass production process, the useful life of certain pieces of equipment. The result of a single observation of these, and many other, phenomena is unpredictably irregular while one notices in a great number of repetitions a certain uniformity. One says that these observations or experiments show statistical regularity. The aim of probability theory is the construction of a mathematical model for such phenomena, called random phenomena. To construct such a model one must describe the experiment precisely by listing all possible outcomes which should be reflected in the model. Thus in tossing a coin one is interested only in the results "head" or "tail" but will disregard freak accidents such as the coin standing on its edge or its disappearing in a hole in the ground. Similarly, in throwing one die we have six possible outcomes which we describe by the number which appears on the top face. The results of the experiments are called "events." In the case of throwing a single die one can also bet whether an even or an odd number appears on the top face. The event of obtaining an even number can be realized in three different ways (either 2 or 4 or 6). The event "throwing a 2 with a single die" can occur in only one way and we are led to distinguish between compound events and simple events. Suppose that we are making a large number, say m, throws with one die and obtain a "6" h times. We say then that $\frac{h}{m}$ is the relative frequency in m throws of the event "a six appears." If we are using a die that is not loaded then we will observe that the relative frequency $\frac{h}{m}$ will be close to $\frac{1}{6}$ if m is large. The statement that the event of obtaining a 6 in our experiment shows statistical regularity means that the relative frequency of its outcome has a certain stability. Our model should also reflect this property of events, this will be accomplished by assigning to each event a number, called its probability, which is an idealization of the relative frequency.

References

Gerolamo Cardano (1501–1576), "De ludo aleae." Published 1663 in the first volume of Cardano's collected works. An English translation can be found in Ore.

G. Galilei (1564–1642), "Considerazione sopra il gioco dei daddi" (date unknown). Vol. 3, Opere, Firenze 1718. Vol. 14, Opere, Firenze 1855.

B. Pascal (1623–1662), "Oeuvres," Vol. IV. Paris 1819.

P. Fermat (1601–1665), "Varia opera mathematica." Tolosae 1679.

O. Ore, "Cardano the Gambling Scholar." Princeton Univ. Press, Princeton, New Jersey, 1953. Reprinted by Dover, New York, 1965.

D. Hilbert, Mathematical Problems (lecture presented before the International Congress of Mathematicians. Paris 1900). English translation: *Bull. Amer. Math. Soc.* **8**, 437–479, (1901–1902).

A. N. Kolmogorov, "Foundations of the Theory of Probability." Springer, Berlin 1933. English translation: Chelsea, Bronx, New York, 1956.

H. Weyl, "Philosophy of Mathematics and Natural Science," p. 27. Princeton Univ. Press, Princeton, New Jersey, 1949.

Surveys of the history of probability theory can be found in:

I. Todhunter, "History of the Theory of Probability." Cambridge, 1865. Reprinted by Chelsea, Bronx, New York, 1949, 1965.

E. Czuber, "Die Entwicklung der Wahrscheinlichkeitstheorie und ihrer Anwendungen." Jahresberichte d. Deutschen Mathematiker Vereinigung **VII**, No. 2, 1–127, 1899.

I | PROBABILITY THEORY

1 | The Probability Space

1.1 The Outcome Space

The first step in constructing a model for random phenomena is the listing of all relevant outcomes. We denote the set (collection) of these by Ω and call Ω the outcome space. The simple events are the elements of Ω.

Example 1.1.1

We consider tossing three (distinguishable) coins simultaneously. There are eight possible outcomes†:

---

† H stands for head, T for tail. The first, second and third letter indicate the face observed on the first, second, and third coins respectively.

$$\omega_1 = (HHH), \quad \omega_2 = (HHT), \quad \omega_3 = (HTH), \quad \omega_4 = (HTT),$$
$$\omega_5 = (THH), \quad \omega_6 = (THT), \quad \omega_7 = (TTH), \quad \omega_8 = (TTT).$$

The outcome space $\Omega = (\omega_1, \omega_2, \omega_3, \omega_4, \omega_5, \omega_6, \omega_7, \omega_8)$ consists here of the eight simple events (outcomes), each simple event contains exactly one point of Ω. One can also consider compound events; we list a few examples:

A: All three coins show the same face.
B: The same face appears on exactly two of the three coins.
C: At least two coins show H.
D: Exactly two coins show H.
E: Exactly two coins show T.
F: H appears on the first and second coin.

The compound events can be described by listing the simple events (points of Ω) which they contain; we say that the compound events are subsets of Ω. For instance,

$$A = (\omega_1, \omega_8), \quad\quad B = (\omega_2, \omega_3, \omega_4, \omega_5, \omega_6, \omega_7),$$
$$C = (\omega_1, \omega_2, \omega_3, \omega_5), \quad D = (\omega_2, \omega_3, \omega_5),$$
$$E = (\omega_4, \omega_6, \omega_7), \quad\quad F = (\omega_1, \omega_2).$$

Having defined a certain event we can define the opposite event. Similarly, we can define the event which occurs if at least one of two events is observed or the event which occurs if two events occur simultaneously. To denote these events we use the notations of set theory and write A^c for the opposite (complementary) event, $A \cup B$ (respectively, $A \cap B$) for the event that at least one of the events A and B occurs (respectively that A and B occur simultaneously).† We can also define the difference of two events by $A - B = A \cap B^c$. The outcome

---

† Using the terminology of set theory we say that $A \cup B$ is the union; $A \cap B$ the intersection of A and B. For more details concerning the elements of set theory see [1, 2, or 3].

1.1 THE OUTCOME SPACE

space Ω is an event; it is called the certain event. We also introduce the impossible event $\varnothing$, clearly $\varnothing = \Omega^c$. In our example we have: $B^c = A$, $B \cap C = D$, $D \cup E = B$, $B - D = E$, $D \cap E = \varnothing$. We also see that every point of D (simple event constituting D) occurs in B so that B occurs whenever D occurs. We say that D implies B and write $D \subset B$ (or $B \supset D$).

We formed unions and intersections of pairs of events, these operations can be generalized to any number of events. We use the notations $\bigcup_{j=1}^{n} A_j \left(\bigcap_{j=1}^{n} A_j \right)$ to express the event that at least one of the n events $A_1, A_2, \ldots, A_n$ occurs (all n occur simultaneously). We have seen that compound events correspond to subsets of the sample space so that relations between events can be formulated in the terminology of set theory. Table 1.1 indicates this correspondence.

The set-theoretic operations of forming unions, intersections, and complements obey certain rules which are similar to the rules of algebra. One speaks therefore also of an algebra of events (sets). We list a few of these rules.

$$A \cup B = B \cup A; \quad A \cap B = B \cap A. \tag{1.1.1}$$

$$(A \cup B) \cup C = A \cup (B \cup C); \quad (A \cap B) \cap C = A \cap (B \cap C). \tag{1.1.2}$$

$$(A \cup B) \cap C = (A \cap C) \cup (B \cap C),$$
$$(A \cap B) \cup C = (A \cup C) \cap (B \cup C). \tag{1.1.3}$$

Formulas (1.1.1), (1.1.2), and (1.1.3) are called the commutative, associative, and distributive laws, respectively.

The following relations hold for $\varnothing$ and Ω

$$A \cup \varnothing = A, \quad A \cap \varnothing = \varnothing; \quad A \cup \Omega = \Omega, \quad A \cap \Omega = A. \tag{1.1.4}$$

We note also that

$$A \cup A = A \cap A = A. \tag{1.1.5}$$

For forming complements we have the following rules:

$$(A^c)^c = A, \quad A \cup A^c = \Omega, \quad A \cap A^c = \varnothing, \quad A^c = \Omega - A. \tag{1.1.6}$$

Table 1.1

Corresponding terms for		
Events	Sets	Notation
Simple event	Point of outcome space	$\omega, \omega_1, \omega_2, \ldots$
Compound event	Subset of outcome space	$A, B, \ldots$
Simple event ω is a constituent of the compound event A	Point ω is an element of A	$\omega \in A$
The certain event	The outcome space	Ω
The impossible event	The empty set	$\emptyset$
The opposite event to A	Complement of A	A^c
The event "A or B" (at least one)	Union of A and B	$A \cup B$
The event "A and B" (both simultaneously)	Intersection of A and B	$A \cap B$
A and B are incompatible (or mutually exclusive)	Intersection of A and B is empty (A and B are disjoint sets)	$A \cap B = \emptyset$
A implies B	A part (subset) of B	$A \subset B$ or $B \supset A$
Events A and B are equivalent	Sets A and B are equal	$A = B$
Event A occurs but not B	Difference of sets A and B	$A - B = A \cap B^c$

The inclusion relation is transitive, this means that $A \subset B$ and $B \subset C$ imply that $A \subset C$. It is also useful to list the relation between the three basic operations of forming unions, intersections and complements.

$$(A \cup B)^c = A^c \cap B^c, \qquad (A \cap B)^c = A^c \cup B^c \qquad (1.1.7)$$

and similarly

$$\left(\bigcup_{j=1}^n A_j\right)^c = \bigcap_{j=1}^n A_j^c, \qquad \left(\bigcap_{j=1}^n A_j\right)^c = \bigcup_{j=1}^n A_j^c.$$

Similarly, one has for an infinite sequence $\{A_j\}$

$$\left(\bigcup_{j=1}^\infty A_j\right)^c = \bigcap_{j=1}^\infty A_j^c; \qquad \left(\bigcap_{j=1}^\infty A_j\right)^c = \bigcup_{j=1}^\infty A_j^c. \qquad (1.1.8)$$

The reader should verify these relations.

A set $\{A_j\}$ of mutually exclusive events is said to be exhaustive if their union is the whole outcome space. This means that every point of the outcome space belongs to one, and only one of the sets A_j.

1.2 Probabilities

We mentioned in the introduction that a model for random phenomena must account for the statistical regularity that these phenomena exhibit. This is done by assigning to each event a number, called its probability. The fact that a fixed number is assigned to each event reflects the stability of the relative frequencies that we observed. In order to make reasonable assumptions concerning the probabilities we discuss first properties of relative frequencies. We consider an experiment concerning a (simple or compound) event A and repeat it m times. Suppose that A occurs in h of the m trials. Clearly

$$0 \leq h \leq m.$$

Let

$$R(A) = \frac{h}{m}$$

be the relative frequency of A in m trials, then

$$0 \leq R(A) \leq 1. \qquad (1.2.1)$$

Consider next two mutually exclusive events A and B (that is, $A \cap B = \emptyset$) and suppose that in m trials A occurs h_A times while B occurs h_B times. Let $C = A \cup B$ be the event that at least one of the two events A or B occurs. It is then easily seen that

$$R(C) = R(A \cup B) = R(A) + R(B) \quad \text{if } A \cap B = \emptyset. \qquad (1.2.2)$$

It is clear that (1.2.2) can be generalized to any finite number of mutually exclusive events.

We consider next an experiment with a finite number of relevant outcomes; these are, of course, mutually exclusive. Let n be the number of possible outcomes and suppose that we observed the jth outcome in m trials h_j times. Then $\sum_{j=1}^{n} h_j = m$ and we see that $\sum_{j=1}^{n} R_j = 1$. Here, R_j is the relative frequency of the jth outcome.

Since the experiment has n relevant outcomes, we select a sample space of n points, $\Omega = (\omega_1, \omega_2, \ldots, \omega_n)$. To each point ω_j we assign a number, its probability $P(\omega_j) = p_j$ which satisfies the following conditions:

$$0 \le p_j \le 1, \tag{1.2.3}$$

$$\sum_{j=1}^{n} p_j = 1. \tag{1.2.4}$$

These conditions reflect the properties (1.2.1) and (1.2.2) of relative frequencies. Let A be a compound event, we assign to it the probability

$$P(A) = \sum_{\omega_j \in A} p_j, \tag{1.2.5}$$

this means that the probability of a compound event equals the sum of the probabilities of the simple events which are its elements.

We consider next experiments whose outcomes form a denumerable† set. As an example we mention the tossing of a coin until head appears for the first time. The first "head" can occur after n tosses, where n is an arbitrary natural number. The outcome space, appropriate to such an experiment, is a denumerable set of points ω_j ($j = 1, 2, \ldots$, *ad infinitum*); the assignment of probabilities to its points can be carried

---

† A set is said to be denumerable if its elements can be put into a one-to-one correspondence with the set of all integers. A set is said to be countable if it is either finite or denumerable. It is known that the set of all rational numbers is denumerable while the set of all real numbers is not countable.

1.2 PROBABILITIES

out in the same way as in the case of a finite sample space. Relation (1.2.3) must hold for all j, while (1.2.4) is replaced by the equation†

$$\sum_{j=1}^{\infty} p_j = 1. \qquad (1.2.4a)$$

Probabilities of compound events are again determined by (1.2.5).

Finite or denumerable outcome spaces are often suitable in model building, however there are experiments whose outcomes do not form a finite or a denumerable set. We turn to the construction of a model for such experiments and discuss first a simple example.

Example 1.2.1

We consider a pointer which is free to rotate about the center of a circle (see Figure 1.1). The pointer is given an initial impulse and will

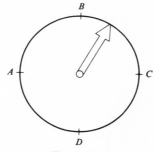

Figure 1.1

spin around the circle and finally come to rest at some point. We assume that the mechanism imparting the impulse to the pointer is not rigged, that is, the pointer has no tendency to stop more (or less) frequently at certain points. The location at which the pointer comes to rest can be used to make bets. It is natural to use the perimeter of the

---

† Equation (1.2.4a) makes it impossible to assign to all points of a denumerable outcome space the same probability.

circle as the outcome space since each of its points is a possible outcome. This sample space is a nondenumerable set. The event that the pointer will stop at a specified point is a simple event. In view of our assumption concerning the mechanism, we must assign the same probability to each point. Since the outcome space is nondenumerable, we cannot assign equal positive probabilities to its points without violating the condition that the sum of all probabilities equals unity. The subsets of the perimeter are the compound events of our outcome space. One could also bet that the pointer stops at a point belonging to a subset, for instance at a point of the arc AB of one of the quadrants. In view of our assumption concerning the mechanism of the pointer, we expect that the relative frequency of this event will be close to $\frac{1}{4}$ so that it is reasonable to assign to it the probability $\frac{1}{4}$.

We see therefore that we cannot expect that positive probabilities can be assigned to the simple events of a noncountable sample space; however, it is possible to assign positive probabilities to compound events. In more technical language, this means that probability cannot be defined as a point function, but must be defined as a set function if the outcome space is noncountable. Whereas one could assign probabilities to all subsets of a countable outcome space, this is not the case for a noncountable outcome space. It can be shown that it is not possible to assign probabilities to all subsets of a general (noncountable) outcome space. There will be a system of subsets to which probabilities are assigned and Section 1.3 will take this into account.

1.3 The Axioms

We now leave the heuristic considerations of the preceding sections and introduce a formal, axiomatic foundation of probability theory. The approach that we present here is due to Kolmogorov.

Suppose that a set Ω and a system $\mathfrak{U}$ of subsets of Ω is given. Ω is called the outcome space and the sets of $\mathfrak{U}$ are called the events. The set Ω and the system $\mathfrak{U}$ satisfy the following conditions:

1.3 THE AXIOMS

(IA) $\Omega \in \mathfrak{U}$,
(IB)   If $A \in \mathfrak{U}$, then $A^c \in \mathfrak{U}$,
(IC)   Let $\{A_j\}$ be a sequence of sets and suppose that $A_j \in \mathfrak{U}$ for all j, then $\bigcup_{j=1}^{\infty} A_j \in \mathfrak{U}$.

A system of subsets which satisfies these conditions is called a σ-algebra (or a σ-field).

We introduce a set function $P(A)$, defined for all $A \in \mathfrak{U}$. This function is called the probability of the event A and satisfies the following conditions:

(IIA) $P(A) \geq 0$ for all $A \in \mathfrak{U}$,
(IIB) $P(\Omega) = 1$
(IIC)   Let $\{A_j\}$ be a sequence of events such that $A_j \cap A_k = \emptyset$ for $j \neq k$ then $P\left(\bigcup_{j=1}^{\infty} A_j\right) = \sum_{j=1}^{\infty} P(A_j)$.

A set function that has the property (IIC) is said to be completely additive.

The triple $(\Omega, \mathfrak{U}, P)$ consisting of a set Ω, a σ-field $\mathfrak{U}$ of its subsets, and a nonnegative, completely additive set function $P(A)$, which is normed so as to assign the value 1 to Ω, is called a probability space.

Example 1.3.1

As a first example of a probability space, we consider a finite outcome space Ω. Let $\Omega = (\omega_1, \omega_2, \ldots, \omega_n)$ be a set containing n points. The set Ω has 2^n subsets, these include Ω itself, the empty set $\emptyset$ as well as the n subsets $\{\omega_j\}$ ($j = 1, \ldots, n$) where $\{\omega_j\}$ contains only the single point $\omega_j \in \Omega$. The system of subsets of Ω is denoted by $\mathfrak{P}(\Omega)$, it is easily seen that $\mathfrak{P}(\Omega)$ is a σ-field, we set $\mathfrak{U} = \mathfrak{P}(\Omega)$. Let $A \in \mathfrak{U}$ and suppose that A contains k of the n points of Ω, we define then $P(A) = \dfrac{k}{n}$. The triple $(\Omega, \mathfrak{U}, P)$ is then a (finite) probability space. The determination of probability that we have just described is called the assignment of uniform probabilities since it assigns to each $\{\omega_j\} \in \mathfrak{U}$ the same probability $\frac{1}{n}$.

More generally, one can assign to each $\{\omega_j\}$ a number $p(\omega_j)$ such that $0 \le p(\omega_j) \le 1$ and $\sum_{j=1}^{n} p(\omega_j) = 1$ and define $P(A) = \sum_{\omega_j \in A} p(\omega_j)$, then $(\Omega, \mathfrak{P}(\Omega), P)$ is again a finite probability space. The construction of this probability space agrees with formulas (1.2.3), (1.2.4), and (1.2.5).

Example 1.3.2

Suppose that the outcome space Ω is denumerable, $\Omega = (\omega_1, \omega_2, \ldots,$ *ad infinitum*). In this case it is impossible to assign uniform probabilities; however one can still construct a probability space by assigning to each $\{\omega_j\}$ a number $p(\omega_j)$ satisfying the conditions $0 \le p(\omega_j) \le 1$ and $\sum_{j=1}^{\infty} p(\omega_j) = 1$. If one defines $P(A) = \sum_{\omega_j \in A} p(\omega_j)$ one obtains the denumerable probability space $(\Omega, \mathfrak{P}(\Omega), P)$.

1.4 Problems

1. We introduced the operations of union and intersection of events in Section 1.1. Prove that these operations satisfy the commutative, associative, and distributive laws.
2. Let A and B be two events. Show that (*a*) $(A \cup B)^c = A^c \cap B^c$ and (*b*) $(A \cap B)^c = A^c \cup B^c$.
3. Show that the events A, $A^c \cap B$, and $(A \cup B)^c$ form an exhaustive set of events.
4. Form the sets $A \cup B$, $A \cap B$, A^c, and $A \cap B^c$ if A is the set of integers exceeding four but less than 21 and B the set of integers exceeding 11 but less than 26.
5. A committee of five members decides on a motion by secret ballot. Abstentions are permitted. Describe the sets corresponding to the following outcomes. (*a*) The motion is carried. (*b*) A tie occurs.

1.4 PROBLEMS

6. In tossing two coins simultaneously, we denote by A the event that head appears exactly once. What is the event A^c?
7. Show that $(A \cup B) - B = A \cap B^c$.
8. Show that $A^c = B^c$ implies that $A = B$.
9. Show that $(A - B) \cup (B - A) = (A \cup B) - (A \cap B)$.
10. Show that $(A - B) \cap B = \emptyset$.
11. Show that $(A \cup B) \cap (A \cup C) = A \cup (B \cap C)$.
12. Show that $A^c \cap (B \cap C)^c = (A^c \cap B^c) \cup (A^c \cap C^c)$.
13. Suppose that $A \supset B$ and $B \supset C$ and show that $A \supset C$.
14. Find a simple expression for $(A \cup B) \cap (A \cup B^c)$.
15. What is the meaning of the relation $A \cap B = A$?
16. Show that $(A^c)^c = A$.
17. Describe a probability space which corresponds to the simultaneous tossing of two unbiased† coins.
18. Construct a probability space for the experiment of throwing an unloaded (that is, unbiased) die.
19. Construct a probability space for simultaneously throwing two unloaded dice.
20. Suppose that you throw a die which is loaded so as to make the appearance of 2, 3, 4, 5 equally likely while the 6 has twice the chance to appear than the 1. Construct a probability space for this game.
21. What is the assignment of probabilities in Problem 20 if one assumes that at least one of the faces 2, 3, 4, 5 will appear in the long run in approximately half of the trials?
22. A die has the form of an irregular tetrahedron. The probability of the outcome of face I, II, III is 0.24, 0.25, and 0.23, respectively. What is the probability of face IV?

---

† We talk about unbiased or true coins (or dice) to indicate that the assignment of uniform probabilities is justified.

23. A number is selected at random† from the numbers 1 to 10.
  (*a*) Let A be the event that the selected number is even and let B be the event that it is a prime. What is the event $A \cap B$ and what is its probability?
  (*b*) Let C be the event that the number selected is divisible by 3. What is the event $A - C$ and what is its probability?

References

1. Robert R. Stoll, "Sets, Logic and Axiomatic Theories." Freeman, San Francisco, 1961.
2. E. Kamke, "Theory of Sets." Dover, New York, 1950.
3. P. R. Halmos, "Naive Set Theory," Van Nostrand-Reinhold Princeton, New Jersey, 1960.

---

† The phrases "drawn at random" or "selected at random" are often used to indicate that uniform probabilities are to be assigned to the outcome space describing the experiment.

2 | Elementary Properties of Probability Spaces

2.1 Simple Consequences of the Axioms

We derive a few consequences of (IA), (IB), and (IC). Since $\emptyset = \Omega^c$ we conclude from (IA) and (IB) that

$$\emptyset \in \mathfrak{U}. \tag{2.1.1}$$

Let $A_1, A_2, \ldots, A_n$ be n events, putting $A_j = A_n$ ($j = n+1, n+2, \ldots$, ad infinitum), we see from (IC) and (1.1.5) that

$$\bigcup_{j=1}^{n} A_j \in \mathfrak{U} \tag{2.1.2}$$

so that every finite union of sets from $\mathfrak{U}$ belongs to $\mathfrak{U}$. Let $\{A_j\}$ be a countable sequence of events, then one obtains from (1.1.8) and (IC) the relation

$$\bigcap_j A_j \in \mathfrak{U}. \tag{2.1.3}$$

Let $n > 1$ and suppose that the $\{A_j\}$ are a sequence of mutually exclusive events. It follows from (IIC) and (IIA) that

$$P\left(\bigcup_{j=1}^{\infty} A_j\right) = \sum_{j=1}^{\infty} P(A_j) \geq \sum_{j=1}^{n} P(A_j).$$

Choose now $A_j = \emptyset$ for all j, then

$$P(\emptyset) \geq nP(\emptyset),$$

and we conclude

$$P(\emptyset) = 0. \tag{2.1.4}$$

Relation (2.1.4) can be used to show that the probability is also finitely additive, this means that for a finite set $A_1, A_2, \ldots, A_n$ of mutually exclusive events one has

$$P\left(\bigcup_{j=1}^{n} A_j\right) = \sum_{j=1}^{n} P(A_j). \tag{2.1.5}$$

Equation (2.1.5) follows from (2.1.4) and (IIC). A set function that is completely as well as finitely additive is said to be countably additive.

We note that $A \cup A^c = \Omega$ and conclude from (IIB) and (2.1.5) that

$$P(A) = 1 - P(A^c). \tag{2.1.6}$$

Since $P(A^c) \geq 0$ [see (IIA)] we see that

$$P(A) \leq 1. \tag{2.1.7}$$

Let A and B be two (not necessarily incompatible) events, then

$$A \cup B = A \cup (B - A),$$

2.1 SIMPLE CONSEQUENCES OF THE AXIOMS

and
$$(A \cap B) \cup (B - A) = B.$$

The events $A \cap B$ and $B - A$, as well as the events A and $B - A$ are mutually exclusive and we see from (2.1.5) that

$$\begin{aligned} P(A \cup B) &= P(A) + P(B - A), \\ P(A \cap B) + P(B - A) &= P(B). \end{aligned} \qquad (2.1.8)$$

We eliminate $P(B - A)$ from these two equations and obtain

Theorem 2.1.1 (The Addition Rule)

For any two events A and B one has

$$P(A \cup B) = P(A) + P(B) - P(A \cap B).$$

Corollary to Theorem 2.1.1

For any two events A and B one has

$$P(A \cup B) \le P(A) + P(B).$$

The corollary can be extended to an arbitrary number of events

$$P\left(\bigcup_j A_j\right) \le \sum_j P(A_j). \qquad (2.1.9)$$

Next we derive some useful inequalities. Let A and B be two events, we see from the addition rule that

$$P(A) + P(B) = P(A \cup B) + P(A \cap B),$$

hence
$$P(A \cap B) = P(A) + 1 - P(B^c) - P(A \cup B),$$
or, in view of (2.1.7)
$$P(A \cap B) \ge P(A) - P(B^c) = 1 - P(A^c) - P(B^c). \qquad (2.1.10)$$

Inequality (2.1.10) is called Boole's inequality. Boole's inequality can easily be extended to an arbitrary number of events.

Theorem 2.1.2 (Boole's Generalized Inequality)

Let $\{A_j\}$ ($j = 1, 2 \ldots$) be a finite or denumerable sequence of events, then the inequality $P(\bigcap_j A_j) \geq 1 - \sum_j P(A_j^c)$ holds.

First we consider the case of a denumerable sequence and put $B_1 = A_1$, $B_2 = \bigcap_{j=2}^{\infty} A_j$ so that $\bigcap_{j=1}^{\infty} A_j = B_1 \cap B_2$. It follows from (2.1.10) that

$$P\left(\bigcap_{j=1}^{\infty} A_j\right) \geq 1 - P(A_1^c) - P(B_2^c).$$

Since

$$P(B_2^c) = P\left(\bigcup_{j=2}^{\infty} A_j^c\right) \leq \sum_{j=2}^{\infty} P(A_j^c),$$

we see that

$$P\left(\bigcap_{j=1}^{\infty} A_j\right) \geq 1 - \sum_{j=1}^{\infty} P(A_j^c),$$

which is the statement of the theorem. If one has only a finite number, say n, events, then one puts $A_j = \Omega$ for $j > n$ and obtains the desired result.

Let A and B be two events and suppose that A implies B; that is $A \subset B$. Then $A \cup B = B$ and we see from the first equation (2.1.8) that

$$P(B) \geq P(A) \quad \text{provided that} \quad A \subset B. \tag{2.1.11}$$

Theorem 2.1.3 (The Implication Rule)

Let A, B, and C be three events, and assume that the simultaneous occurrence of A and B implies C, then $P(C^c) \leq P(A^c) + P(B^c)$.

2.1 SIMPLE CONSEQUENCES OF THE AXIOMS

To prove the theorem we rewrite Boole's inequality in the form

$$P(A^c) + P(B^c) \geq 1 - P(A \cap B) = P(A^c \cup B^c).$$

According to the assumption of the theorem $A \cap B \subset C$ so that $A^c \cup B^c \supset C^c$. The statement of the theorem then follows immediately from (2.1.11).

Sometimes it is necessary to consider infinite sequences of events $\{A_n\}$ such that either

$$A_1 \subset A_2 \subset \cdots \subset A_j \subset A_{j+1} \subset \cdots,$$

or

$$A_1 \supset A_2 \supset \cdots \supset A_j \supset A_{j+1} \supset \cdots.$$

In the first case, we say that the sequence is increasing, in the second that it is decreasing. We prove the following statements.

Theorem 2.1.4

Let $\{A_n\}$ be an increasing sequence of events, and let $A = \bigcup_{j=1}^{\infty} A_j$, then $P(A) = \lim_{n \to \infty} P(A_n)$.

Theorem 2.1.5

Let $\{A_n\}$ be a decreasing sequence of events, and let $A = \bigcap_{j=1}^{\infty} A_j$, then $P(A) = \lim_{n \to \infty} P(A_n)$.

Proof of Theorem 2.1.4

We write A as the union of disjoint events,

$$A = A_1 \cup \bigcup_{j=1}^{\infty} (A_{j+1} - A_j),$$

and conclude from the countable additivity of probability [Axiom (IIC)] that

$$P(A) = P(A_1) + \sum_{j=1}^{\infty} P(A_{j+1} - A_j),$$

or

$$P(A) = P(A_1) + \lim_{n \to \infty} \sum_{j=1}^{n-1} [P(A_{j+1}) - P(A_j)]$$

so that $P(A) = \lim_{n \to \infty} P(A_n)$ as stated in the theorem.

Proof of Theorem 2.1.5

Since the $\{A_n\}$ form a decreasing sequence, we see that $\{A_n^c\}$ is an increasing sequence of events and

$$A^c = \bigcup_{j=1}^{\infty} A_j^c.$$

We apply Theorem 2.1.4 and see that

$$P(A^c) = \lim_{n \to \infty} P(A_n^c).$$

The statement of the theorem then follows immediately.

2.2 Conditional Probability and Independence

Sometimes one encounters a situation where one has some incomplete information concerning the outcome of an experiment. More specifically, one knows that a certain event B has occurred and wishes to take this information into account in making a statement concerning the outcome of another event A.

We let ourselves again be guided by the properties of relative frequencies. Suppose that n trials were made and that the events B and

2.2 CONDITIONAL PROBABILITY

$A \cap B$ were observed in h_B and h_{AB} trials, respectively. The relative frequencies of the events are

$$R(B) = \frac{h_B}{n} \quad \text{and} \quad R(A \cap B) = \frac{h_{AB}}{n}.$$

The relative frequency of the event A among the outcomes resulting in B is given by

$$R(A|B) = \frac{h_{AB}}{h_B},$$

and is called the relative frequency of A given B. Clearly $R(A|B)$ is the quotient of two relative frequencies

$$R(A|B) = \frac{R(A \cap B)}{R(B)}.$$

These considerations suggest the following definition.

Let $(\Omega, \mathfrak{U}, P)$ be a probability space and let A and B be two events. Suppose that $P(B) > 0$, the quotient

$$P(A|B) = \frac{P(A \cap B)}{P(B)} \tag{2.2.1}$$

is called the conditional probability of the event A, given that the event B has occurred.

Before discussing properties of conditional probabilities we give an example.

Example 2.2.1

We consider, as in Example 1.1.1, the simultaneous tossing of three distinguishable coins and employ the notation of Example 1.1.1 for the outcome space Ω and for the eight simple events. We construct for this experiment a finite probability space with uniform probabilities (see the construction given in Example 1.3.1). We consider the event $F = (\omega_1, \omega_2)$ that head appears on the first two coins and the event $D = (\omega_2, \omega_3, \omega_5)$

that exactly two coins show head. Then $F \cap D = (\omega_2)$ and $P(F \cap D) = \frac{1}{8}$ while $P(D) = \frac{3}{8}$ so that the conditional probability of head appearing on the first two coins, given that exactly two heads appeared in the throw is $P(F|D) = \frac{1}{3}$.

Theorem 2.2.1

Let B be a fixed event such that $P(B) > 0$, then $(\Omega, \mathfrak{U}, P(\cdot|B))$ is a probability space.

Proof

To prove the theorem we must only show that, for fixed B, $P(A|B)$ satisfies Axioms (IIA), (IIB), and (IIC) for all $A \in \mathfrak{U}$. It follows from (2.2.1) that (IIA) is satisfied, (IIB) is also valid since

$$P(\Omega|B) = \frac{P(\Omega \cap B)}{P(B)} = \frac{P(B)}{P(B)} = 1.$$

We must still show that $P(A|B)$ is countably additive. Let $\{A_j\}$ be a sequence of mutually exclusive events then

$$P\left(\bigcup_{j=1}^{\infty} A_j \mid B\right) = \frac{P(\bigcup_{j=1}^{\infty} B \cap A_j)}{P(B)} = \frac{[\sum_{j=1}^{\infty} P(A_j \cap B)]}{P(B)} = \sum_{j=1}^{\infty} P(A_j|B).$$

This completes the proof of the theorem.

Theorem 2.2.1 indicates that conditional probabilities have all the properties that we proved for ordinary (unconditional) probabilities. We now derive some theorems that form a link between conditional and ordinary probabilities.

Let A and B be two events and suppose that $P(A) > 0$ and $P(B) > 0$. We see then from (2.2.1) that

$$P(A \cap B) = P(B)P(A|B),$$
$$P(A \cap B) = P(A)P(B|A).$$

2.2 CONDITIONAL PROBABILITY

We eliminate $P(A \cap B)$ and obtain

$$P(A|B) = \frac{P(A)P(B|A)}{P(B)}. \tag{2.2.2}$$

Formula (2.2.2) is sometimes called Bayes' formula.

Let $A_1, A_2, \ldots, A_n$ be n events and assume that

$$P\left(\bigcap_{j=1}^{n-1} A_j\right) > 0. \tag{2.2.3}$$

Since $\bigcap_{j=1}^{n-1} A_j \subset \bigcap_{j=1}^{n-2} A_j \subset \cdots \subset (A_2 \cap A_1) \subset A_1$, we conclude from (2.2.3) that $P\left(\bigcap_{j=1}^{n-k} A_j\right) > 0$ for $k = 1, 2, \ldots, n-1$. The conditional probabilities $P\left(A_k \middle| \bigcap_{j=1}^{k-1} A_j\right)$ are therefore defined and we obtain the following result.

Theorem 2.2.2 (The Multiplication Rule)

Let $A_1, A_2, \ldots, A_n$ be n events and assume that $P\left(\bigcap_{j=1}^{n-1} A_j\right) > 0$. Then

$$P\left(\bigcap_{j=1}^{n} A_j\right) = P(A_1)P(A_2|A_1)P(A_3|A_1 \cap A_2) \cdots P\left(A_n \middle| \bigcap_{j=1}^{n-1} A_j\right).$$

Another result is also very important.

Theorem 2.2.3 (The Total Probability Rule)

Let $\{A_j\}$ be a countable set of exhaustive events, and suppose that $P(A_j) > 0$. Let B be an arbitrary event, then

$$P(B) = \sum_j P(A_j)P(B|A_j).$$

According to the assumptions of the theorem, one has $\bigcup_j A_j = \Omega$ and

$A_j \cap A_k = \emptyset$ for $j \neq k$. Therefore $B = \bigcup_j A_j \cap B$ and $P(B) = \sum_j P(A_j \cap B)$. The statement of the theorem follows from (2.2.1).

Next, we prove a simple consequence of this theorem.

Theorem 2.2.4 (Bayes' Theorem)

Let $\{A_j\}$ be a countable set of exhaustive events such that $P(A_j) > 0$ and let B be an event with $P(B) > 0$. Then the relation

$$P(A_j|B) = \frac{P(A_j)P(B|A_j)}{\sum_j P(A_j)P(B|A_j)}$$

holds for all j.

To prove Theorem 2.2.4, we note that according to (2.2.2), one has

$$P(A_j|B) = \frac{P(A_j)P(B|A_j)}{P(B)}.$$

If one applies the total probability rule to $P(B)$, one obtains the desired result.

Theorem 2.2.4 is sometimes called "Bayes' rule for the probability of causes." The reason for this name is that the theorem has the following interpretation: Assume that the occurrence of any event A_j of the exhaustive set $\{A_j\}$ has an effect on the event B. Then one can regard the A_j as "causes" for the event B, and interpret the conditional probability $P(A_j|B)$ as the probability that the observed event B was caused by the event A_j. This probability can be computed from Bayes' theorem, provided that (*i*) the *a priori* probabilities $P(A_j)$ and (*ii*) the conditional probabilities $P(B|A_j)$ are given. If Bayes' rule is applied when these two conditions are not satisfied then the conclusions will be invalid.

The introduction of conditional probabilities was motivated by the fact that information concerning a certain event B may affect our appraisal of the probability of another event A. However, it can also

2.2 CONDITIONAL PROBABILITY

happen that the occurrence of an event B does not influence the outcome of an experiment concerning a second event A. This last possibility is very important in probability theory and mathematical statistics and is formalized in the following manner:

Two events A and B are said to be independent if $P(A \cap B) = P(A)P(B)$.

Theorem 2.2.5

Let A and B be two independent events, and assume that $P(B) > 0$, then $P(A|B) = P(A)$.

This follows immediately from formula (2.2.1); the theorem expresses our intuitive idea of independence, namely that the outcome of B does not affect A.

Example 2.2.1 and Example 1.1.1 (continued)

We consider in this example the events G_1: "H appears on the first coin" and G_2: "H appears on the second coin." Clearly, $G_1 = (\omega_1, \omega_2, \omega_3, \omega_4)$, whereas $G_2 = (\omega_1, \omega_2, \omega_5, \omega_6)$. Then $G_1 \cap G_2 = (\omega_1, \omega_2)$ is the event F that H appears on the first as well as on the second coin. It is easily seen that $P(G_1) = P(G_2) = \frac{1}{2}$ while $P(G_1 \cap G_2) = \frac{1}{4}$ so that G_1 and G_2 are independent.

Theorem 2.2.6

Let A and B be two mutually exclusive (incompatible) events and suppose that $P(A) > 0$, $P(B) > 0$, then A and B cannot be independent.

It follows that $A \cap B = \emptyset$ hence $P(A \cap B) = 0$, whereas $P(A)P(B) > 0$. This precludes independence.

The concept of independence can be extended to an arbitrary number of events. We consider an example that will give us some indication of how to proceed.

Example 2.2.2

We consider the simultaneous tossing of two distinguishable, unbiased coins, say a nickel and a dime. The outcome space has four points and we again assign uniform probabilities. Let A be the event "H on the nickel," B be the event "H on the dime," and let C be the event that one of the two coins shows H while the other shows T. The event $A \cap B$ means that H appears on both coins, $A \cap C$ means that H is on the nickel while T is on the dime, and $B \cap C$ means T on the nickel H on the dime.

It is then easily seen that $P(A) = P(B) = P(C) = \frac{1}{2}$ while $P(A \cap B) = P(A \cap C) = P(B \cap C) = \frac{1}{4}$. We see, therefore, that the three pairs of events (A, B), (A, C), and (B, C) are pairs of independent events. However $A \cap B$ and C are incompatible. It would, therefore, seem inappropriate to call the three events A, B, C independent.

This situation motivates the following definition:

The n events $A_1, A_2, \ldots, A_n$ are said to be (completely) independent† if for any selection $A_{i_1}, A_{i_2}, \ldots, A_{i_k}$ of k of the n events ($1 \leq i_1 < i_2 \cdots < i_k \leq n$, $k = 2, 3, \ldots, n$) the $2^n - n - 1$ relations

$$P\left(\bigcap_{j=1}^{k} A_{i_j}\right) = \prod_{j=1}^{k} P(A_{i_j}) \qquad (2.2.4)$$

hold.

Formula (2.2.4) means that n events are independent if the probability of the joint occurrence of any finite number of them is equal to the product of their probabilities.

The definition of independence can be extended to arbitrary (nondenumerable) sets of events.

---

† In general, we shall omit the adjective "completely" and talk simply about independent events. However, we shall use the term "pairwise independent" if only any two of the n events are independent.

2.3 Finite Probability Spaces

Finite probability spaces were introduced in Example 1.3.1. The simplest cases are finite probability spaces with uniform probabilities. These were the first probability spaces studied; they are completely determined if the simple events are known, since the probability of any event A equals the ratio $\frac{k}{n}$, where n is the total number of simple events whereas k is the number of simple events contained in A (called outcomes favorable to A). The determination of the probability of an event A reduces then to the problem of counting the outcomes contained in A. Combinatorial methods are often useful in carrying out this task.

Finite probability spaces with uniform probabilities occur in connection with gambling problems, such as tossing coins or dice, drawing at random from urns or from a deck of cards.

As a simple example of a situation in which such a model can be applied, we mention repeated trials of the same experiment, where each trial has only two possible outcomes,† and where the trials are independent. Consider for instance the experiment of tossing an unbiased coin n times. There are 2^n points in the outcome space (possible outcomes). Let A be the event that head appears k times in n tosses, then there are $\binom{n}{k}$ favorable events‡ (corresponding to the ways in which k heads can appear in n tosses). Therefore,

$$P(A) = \frac{\binom{n}{k}}{2^n}$$

is the probability of obtaining exactly k heads in n trials.

---

† Trials that have only two possible outcomes are called (simple) alternatives.

‡ The meaning of the symbol $\binom{n}{k}$ and of related symbols is explained in Appendix A. Readers not familiar with these notations should read Appendix A before proceeding.

A more general model is obtained if one considers the same coin-tossing experiment, but uses a biased coin. Suppose that the probability of head is $p \neq \tfrac{1}{2}$, and the probability of tail is $q = 1 - p$. It is then impossible to assign uniform probabilities, since the probability of obtaining n heads (p^n) is different from the probability of tossing n tails (q^n). The outcome space has still 2^n points and $\binom{n}{k}$ of these are favorable to A, but one must assign to each of these points the probability $p^k q^{n-k}$, so that $P(A) = \binom{n}{k} p^k q^{n-k}$. Repeated independent trials of alternatives occur in many situations,† it is therefore desirable to introduce a neutral terminology and to call the outcomes "success" or "failure."

Theorem 2.3.1

Let S_n be the number of successes in n independent trials of the same alternative, then $P(S_n = k) = \binom{n}{k} p^k q^{n-k}$ for $k = 0, 1, 2, \ldots, n$. Here, p is the probability of success in a single trial while $q = 1 - p$.

The probabilities $P(S_n = k)$ are terms of the binomial expansion of $(p + q)^n$ and are therefore called binomial probabilities. A sequence of trials satisfying the conditions of Theorem 2.3.1 is often called a sequence of Bernoulli trials.

Next, we describe an urn scheme, which leads also to binomial probabilities. An urn contains N balls, M black and $N - M$ white. The experiment consists of drawing consecutively n balls, each ball

---

† As a technical application, we mention the control of a mass production process. The manufactured product is classified to be either good or defective, and it can often be assumed that the probability of producing a defective item is constant. The quality control engineer will not test all manufactured items, but only a certain number of them and will estimate, on the basis of his observation, the total fraction of defective items.

2.3 FINITE PROBABILITY SPACES

is returned to the urn before the next ball is drawn. The balls are well mixed before each drawing so that the probability of drawing a black ball remains unchanged and equals $p = \dfrac{M}{N}$. The probability of drawing k black balls in n drawings is then again a binomial probabilty. This scheme is called "sampling with replacement."

Another urn scheme, called "sampling without replacement" is also often useful. The urn contains as before M black and $N - M$ white balls. Again, n balls are drawn consecutively but the balls are not returned after each drawing. The fraction of black balls in the urn changes therefore with each drawing, the probability of getting a black ball in a particular drawing depends on the results of the previous drawings so that consecutive trials are not independent. For example, if the first drawing resulted in a black ball then the probability of obtaining a black ball in the second trial is $\dfrac{M-1}{N-1}$ while it is $\dfrac{M}{N-1}$ if the first trial resulted in a white ball.

The outcome space for this experiment contains $\binom{N}{n}$ points, and we assign again uniform probabilities. The probability of each simple event is $\dfrac{1}{\binom{N}{n}}$. There are $\binom{M}{k}\binom{N-M}{n-k}$ simple events containing k black balls, since one can select k black balls in $\binom{M}{k}$ different ways, and has then still $\binom{N-M}{n-k}$ ways to select the white balls. Let S_n be the number of black balls selected in n drawings, we obtain

$$P(S_n = k) = \frac{\binom{M}{k}\binom{N-M}{n-k}}{\binom{N}{n}} \qquad (2.3.1)$$

for the probability of drawing k black balls in n drawings without replacement.

2.4 Problems

1. Prove that $P(A \cap B) = P(A) - P(A - B)$.
2. Show that
$$P(A \cup B \cup C) = P(A) + P(B) + P(C) - P(A \cap B) \\ - P(A \cap C) - P(B \cap C) + P(A \cap B \cap C).$$
3. Let A and B be two incompatible events and let $P(A)$ and $P(B)$, respectively, be their probabilities. Let C be the event that neither of the events A and B occurs and give a formula for $P(C)$.
4. Show that the probability that exactly one of the two events A and B occurs is equal to $P(A) + P(B) - 2P(A \cap B)$.
5. Let A and B be two events, show that $P(A - B) = P(A) - P(A \cap B)$.
6. An unbiased die, shaped like a regular octahedron has two of its sides painted black, two white, two red, and two yellow. The die is rolled twice, describe an outcome space for this experiment. How would you assign probabilities?
7. A combination lock contains four disks on a common axis. These disks are divided into ten sectors marked consecutively with the digits $0, 1, \ldots, 9$. The lock can be opened only if the disks occupy definite positions with respect to the case of the lock, in this case, the digits form a definite combination. What is the probability that the lock can be opened by randomly choosing a combination of digits?
8. Urn A contains three white balls and one black ball. Urn B contains one white ball and six black balls. One ball is drawn at random from Urn A and is transferred to Urn B without disclosing its color. Then one ball is drawn from Urn B and it is revealed that it was white. Using Bayes' formula find the probability that the transferred ball was black.

2.4 PROBLEMS

9. Let A, B, C be three events and assume that $P(C) > 0$ and that $A \cap B \cap C = \emptyset$. Show that $P(A \cup B | C) = P(A | C) + P(B | C)$.

10. Prove the identity
$$P(A) + P(A^c \cap B) = P(B) + P(B^c \cap A).$$
for two events.

11. Suppose that two unbiased dice are tossed. Determine the conditional probability that their sum is 4, if it is known that the sum is even.

12. Suppose that A and B are independent events. Show that (a) A and B^c, (b) A^c and B, (c) A^c and B^c are also independent.

13. Let A and B be incompatible events. Give a condition that assures that they are also independent.

14. Suppose that three unbiased coins are tossed. What is the conditional probability that all three coins show the same face if it is known that at least two coins show head?

15. Prove for two events A and B the identity
$$P(A^c \cap B^c) = 1 - P(A) - P(B) + P(A \cap B).$$

16. Let $(\Omega, \mathfrak{U}, P)$ be a probability space and let B be a fixed event such that $P(B) > 0$. Let $\mathfrak{U}_B$ be the system of subsets of B that belongs to $\mathfrak{U}$. Show that $(B, \mathfrak{U}_B, P(\cdot | B))$ is a probability space.

17. The experiment that we discuss requires two urns U_1 and U_2 and an unbiased die. The urn U_1 contains four white and six black balls, the urn U_2 contains five white and five black balls. Two faces of the die are painted red and four green. The experiment is carried out in two steps. First the die is tossed to select an urn, if a red face shows, U_1 is selected; if a green face shows U_2 is chosen. Finally, a single ball is drawn from the urn so selected.
(a) Find the probability of drawing a white ball;
(b) Suppose that it is known that a white ball was drawn, what is the probability that it came from U_1.

18. Suppose that two distinguishable unbiased dice are thrown. Find the conditional probability that the sum of the dots on the two dice is 7 if it is known that their difference is 1.

19. Consider the game of tossing a biased coin until head appears for the first time. Find the probability that the game terminates with the nth toss.
20. Describe the probability space for Problem 19.
21. Let $P(S_n = k)$, $k = 0, 1, \ldots, n$ be the hypergeometric probabilities given by (2.3.1). Show that $\sum_{k=0}^{n} P(S_n = k) = 1$.
22. Consider the game of tossing a biased coin and find the probability that head appears for the first time in an even numbered toss.
23. Consider the game of tossing a biased coin and find the probability that head appears for the first time in an odd numbered toss.
24. In poker each player is dealt five cards. What is the number of possible hands? (A deck of cards has 52 cards.)
25. What is the probability of getting four aces and one king in a hand of poker?
26. Three cards are drawn without replacement from a well-shuffled deck of cards. What is the probability that at least one of the cards is an ace, a ten, or a jack?
27. Determine the probability that in a group of $n \leq 365$ people at least one pair will have a common birthday. How many people do we need to make the probability exceed 50%? (*Note:* Common birthday means same month and same date of the month, take the year to have 365 days. Disregard the possibility of a leap year.)
28. A society of n members $n > 2$ sits down at a round table. What is the probability that two certain persons will sit next to each other?

3 | Random Variables and Their Probability Distributions

3.1 Random Variables

It is frequently natural to associate a number with the outcome of an experiment. This is necessarily the case if the experiment results in counts or in measurements of physical quantities. Games of chance afford also an example, since gains or losses of bets depend on this outcome. The assignment of numbers to the outcome is often convenient even in cases where the outcome is of a qualitative, nonnumerical nature. For instance one can label the outcomes "head" or "tail" in coin tossing by 1 or 0, respectively; more generally one can attach these labels to the outcomes "success" or "failure" of any alternative. If one labels the points of an outcome space, one converts qualitative outcomes into numerical outcomes. In mathematical language, the labeling of the outcomes means that a function† is defined on the outcome space. Clearly, one wishes to make probability statements about these functions; this makes it necessary to impose certain restrictions on the functions that we consider.

We return now to our mathematical model; let $(\Omega, \mathfrak{U}, P)$ be a probability space and let $X(\omega)$ be a function defined on Ω. We introduce the following definition:

A random variable $X = X(\omega)$ is a function defined on Ω which assumes only finite values and has the property that the set of simple events ω for which $X(\omega) \leq x$ is an event for any real x.

We introduce a convenient notation and write $\{\omega : q(\omega)\}$ for the set of all points $\omega \in \Omega$ that satisfy a condition $q(\omega)$. The restriction imposed on a random variable $X(\omega)$ can then be written as

$$\{\omega : X(\omega) \leq x\} \in \mathfrak{U} \qquad \text{for all real} \quad x.$$

This definition of a random variable enables us to introduce the probability that the value of a random variable does not exceed a certain number and we write $P(\{\omega : X(\omega) \leq x\})$ for this probability. We shall denote random variables by capital latin letters and will omit the independent variable ω whenever it will not cause any ambiguity.‡

We note that

$$\bigcup_{n=1}^{\infty} \left[X \leq x - \frac{1}{n} \right] = [X < x], \qquad (3.1.1a)$$

---

† Obviously it would vitiate our purpose if we would attach more than one label to each point of the outcome space. The functions that we discuss will therefore always be single valued and we shall not mention this specifically in the following.

‡ For the sake of simplicity, from now on we shall write $[X(\omega) \leq x]$ or $[X \leq x]$ for the event $\{\omega : X(\omega) \leq x\}$ and $P(X \leq x)$ for its probability. Letting A be a set of real numbers, we shall also use the notation $[X \in A]$ for the set $\{\omega : X(\omega) \in A\}$.

3.1 RANDOM VARIABLES

and

$$\bigcap_{n=1}^{\infty}\left[X < x + \frac{1}{n}\right] = [X \leq x], \tag{3.1.1b}$$

we conclude from (3.1.1a) and the definition of a random variable that $[X < x] \in \mathfrak{U}$. Using also (3.1.1b) one sees easily that the conditions $[X \leq x] \in \mathfrak{U}$ and $[X < x] \in \mathfrak{U}$ are equivalent. Using the properties of the σ-field $\mathfrak{U}$, one sees that the sets $[X > x]$, $[x < X \leq y]$, $[x \leq X \leq y]$, $[x \leq X < y]$, $[X = x]$ belong to $\mathfrak{U}$.

Let X and Y be two random variables defined on the same probability space and let R be the set of all rational numbers. We put

$$A = \bigcup_{r \in R} \{[X < r] \cap [Y < z - r]\}.$$

Since the set of all rational numbers is denumerable we see that

$$A \in \mathfrak{U}. \tag{3.1.2a}$$

Moreover, the event $[X < r] \cap [Y < z - r]$ implies the event $[X + Y < z]$ so that

$$A \subset [X + Y < z]. \tag{3.1.2b}$$

On the other hand, the event $[X + Y < z]$ implies that there exists for each point $\omega \in \Omega$ a rational number r_0 such that $X < r_0$ and $Y < z - r_0$; therefore $[X + Y < z]$ implies the event $[X < r_0] \cap [Y < z - r_0]$ so that

$$[X + Y < z] \subset A. \tag{3.1.2c}$$

We conclude from (3.1.2b) and (3.1.2c) that

$$A = [X + Y < z] \tag{3.1.3}$$

or, in view of (3.1.2a)

$$[X + Y < z] \in \mathfrak{U},$$

so that $X + Y$ is a random variable. A similar reasoning can be applied to $X - Y$ and we obtain the first part of the following theorem:

Theorem 3.1.1

The sum and the difference of two random variables is a random variable. The product of a random variable and a real constant is a random variable.

Let k be a real number, if one considers the set $[kX \leq x]$ for k positive, negative, or zero, one obtains the second part of the statement.

Theorem 3.1.2

The square of a random variable is a random variable. The product of two random variables is a random variable.

Theorem 3.1.3

Let X be a random variable and assume that X cannot assume the value zero. Then $1/X$ is also a random variable.

The proof of Theorems 3.1.2 and 3.1.3 is left as an exercise for the reader.

Corollary to Theorem 3.1.3

Let X and Y be two random variables and assume that Y cannot assume the value zero. The quotient X/Y is then also a random variable.

This corollary follows immediately from Theorems 3.1.2 and 3.1.3.

3.2 Distribution Functions

In this section, we define and discuss distribution functions. These functions are of great importance in probability theory and statistics. It is convenient to introduce certain notations and terms that will be helpful in studying distribution functions.

3.2 DISTRIBUTION FUNCTIONS

Let $g(x)$ be a function, we write

$$g(\infty) = \lim_{x \to \infty} g(x), \qquad g(-\infty) = \lim_{x \to -\infty} g(x),$$

provided that these limits exist.

The function $g(x)$ is said to be continuous to the right (for short, right-continuous) if $\lim_{h \downarrow 0} g(x + h) = g(x)$. Here $h \downarrow 0$ indicates that h tends to zero through positive values (that is, from the right).

Let X be a random variable; the probability $P(X \leq x)$ is a function of the real variable x. We write

$$F_X(x) = P(X \leq x)$$

and call $F_X(x)$ the distribution function† of the random variable X.

Theorem 3.2.1

The distribution function $F_X(x)$ of a random variable X has the following properties:

(i) $F_X(x)$ *is a nondecreasing function of* x.
(ii) $F_X(-\infty) = 0$, $F_X(+\infty) = 1$.
(iii) $F_X(x)$ *is right-continuous.*

Proof

Let $x_1 < x_2$, we note that

$$[X \leq x_2] = [X \leq x_1] \cup [x_1 < X \leq x_2]. \qquad (3.2.1)$$

All the sets occurring in (3.2.1) belong to the σ-field $\mathfrak{U}$ on which P is defined. Therefore we conclude from the finite additivity and nonnegativity of the function $P(A)$ that

$$P(X \leq x_1) \leq P(X \leq x_2)$$

so that (i) is proven.

---

† The subscript X on $F_X(x)$ indicates the random variable whose distribution one considers; it is often possible to omit this subscript without causing any ambiguity. This can be accomplished by using different capital letters of the Latin alphabet for distribution functions of different random variables.

Statement (*ii*) follows easily from Theorems 2.1.4 and 2.1.5 and from the fact that a random variable is, by definition, finite. To prove (*iii*) we consider the sets $A_n = \left[x < X \leq x + \dfrac{1}{n}\right]$ and note that $\bigcap_{n=1}^{\infty} A_n = \varnothing$, it follows then from Theorem 2.1.5 that

$$\lim_{n \to \infty} P(A_n) = P\left(\bigcap_{n=1}^{\infty} A_n\right) = 0.$$

We write this in terms of distribution functions and see that

$$\lim_{n \to \infty} F_X\left(x + \frac{1}{n}\right) = F_X(x)$$

and we conclude then easily that (*iii*) is also valid.

Nondecreasing and right-continuous functions $F(x)$ such that $F(-\infty) = 0$ while $F(+\infty) = 1$ are called distribution functions. The study of distribution functions leads to interesting problems in analysis, their importance for probability theory is indicated by the following theorem, which we state without proof.†

Theorem 3.2.2

Let $F(x)$ be a distribution function, then there exists a probability space $(\Omega, \mathfrak{U}, P)$ and a random variable X defined on it such that $F(x)$ is the distribution function of the random variable X.

A distribution function $F_X(x)$ is always right-continuous, however it need not be continuous so that it is possible that $F_X(x) \neq \lim_{h \downarrow 0} F_X(x - h)$. We write

$$p_x = F_X(x) - \lim_{h \downarrow 0} F_X(x - h) = \lim_{h \downarrow 0} P(x - h < X \leq x)$$

and call p_x the saltus (jump) of $F_X(x)$ at point x.

---

† For a proof of Theorem 3.2.2 see [2, p. 16], and [1, p. 53].

3.2 DISTRIBUTION FUNCTIONS

Let $F(x)$ be the distribution function of a random variable. The distribution function $F(x)$ is continuous at point x if, and only if, $p_x = 0$; we say then that x is a continuity point of $F(x)$. A point x is said to be a discontinuity point of $F(x)$ if the saltus p_x at x is positive. A point x is said to be a point of increase of $F(x)$ if $F(x + \varepsilon) - F(x - \varepsilon) > 0$ for any $\varepsilon > 0$. Every discontinuity point is a point of increase; the converse is not true, a continuity point can also be a point of increase.

We mention without proof† the following result.

Theorem 3.2.3

The only discontinuity points of a distribution function are points at which it has a jump. The set of all discontinuity points of a distribution function is at most denumerable.

We define next certain important classes of distribution functions.

The distribution function $F_X(x)$ of a random variable X is said to be discrete if all points of increase of $F_X(x)$ are discontinuity points.

In this case, there exists a countable (finite or denumerable) set of points x_j such that $P(X = x_j) = p_{x_j} > 0$, with $\sum_j p_{x_j} = 1$. The distribution function of X can then be written as

$$F_X(x) = \sum_{x_j \leq x} p_{x_j}. \tag{3.2.2}$$

The points x_j are called possible values of the discrete random variable X.

The distribution function $F_X(x)$ of a random variable X is said to be continuous if all points of increase of $F_X(x)$ are continuity points.

A subclass of the continuous distributions is of particular interest.

The distribution function $F_X(x)$ of a random variable X is said to be absolutely continuous if there exists a function $p_X(x)$ such that

---

† See [*1*, p. 52].

$F_X(x) = \int_{-\infty}^{x} p_X(y) \, dy$. The function $p_X(y)$ is then called the density function† (frequency function) of the random variable X.

It follows from Theorem 3.2.1 that the frequency function of a random variable X has the following properties.

$$p_X(x) \geq 0 \qquad (3.2.3a)$$

$$\int_{-\infty}^{\infty} p_X(x) \, dx = 1. \qquad (3.2.3b)$$

Conversely, if a function $p(x)$ satisfies (3.2.3a) and (3.2.3b) then $F(x) = \int_{-\infty}^{x} p(y) \, dy$ is a distribution function and we see from Theorem 3.2.2 that there exists a random variable X that has the absolutely continuous distribution $F(x)$ and has therefore $p(x)$ as its frequency function. We also see from the definition of the frequency function that it is the derivative of the corresponding distribution function,

$$p_X(x) = F'_X(x).$$

In this book, we shall deal only with discrete or absolutely continuous‡ distribution functions. However, these are only two disjoint subclasses of distributions. These two classes contain all distributions that are of practical importance in statistics at present.

3.3 Examples of Discrete Distributions

(a) We consider the function $X(\omega) = \xi$, defined on Ω, where ξ is a real constant. Then

$$[X(\omega) \leq x] = \begin{cases} \varnothing & \text{if } x < \xi, \\ \Omega & \text{if } x \geq \xi. \end{cases} \qquad (3.3.1)$$

---

† We shall omit the subscript indicating the random variable whenever it will not cause any ambiguity and write $p(x)$ instead of $p_X(x)$.

‡ Some textbooks of statistics refer to the absolutely continuous distributions simply as continuous distributions. This is, however, an unfortunate and confusing usage, since it can be shown that there exist continuous distributions that are not absolutely continuous. We say that a random variable X is discrete (absolutely continuous) if its distribution function is discrete (absolutely continuous).

3.3 DISCRETE DISTRIBUTIONS

We see therefore that a constant is, according to our definition, a random variable. A random variable which is constant (that is, which can assume only one value) is called an improper or degenerate random variable. We consider the random variable that is 0 at all points of Ω and denote its distribution function by $\varepsilon(x)$. It follows from (3.3.1) that this random variable has the distribution function

$$\varepsilon(x) = \begin{cases} 0 & \text{if } x < 0, \\ 1 & \text{if } x \geq 0. \end{cases} \tag{3.3.2}$$

More generally we see that a random variable which is identically equal to a constant ξ has the distribution function

$$\varepsilon(x - \xi) = \begin{cases} 0 & \text{if } x < \xi, \\ 1 & \text{if } x \geq \xi. \end{cases} \tag{3.3.3}$$

The distribution functions $\varepsilon(x - \xi)$ are called degenerate distributions and are the simplest discrete distributions.

(b) Let $A \in \mathfrak{U}$ be an event. We define the following function on the outcome space:

$$X = X(\omega) = \begin{cases} 1 & \text{if } \omega \in A, \\ 0 & \text{if } \omega \in A^c. \end{cases}$$

Then

$$[X \leq x] = \begin{cases} \Omega & \text{if } x \geq 1, \\ A^c & \text{if } 0 \leq x < 1, \\ \varnothing & \text{if } x < 0. \end{cases}$$

Therefore X is a random variable; it is called the indicator (random variable) of the event and we write I_A for the indicator of a set $A \in \mathfrak{U}$. Indicators describe alternatives, the event A can be regarded as success, A^c as failure of the trial.

The distribution function of the indicator variable I_A is

$$F_{I_A}(x) = \begin{cases} 0 & \text{if } x < 0, \\ P(A^c) & \text{if } 0 \leq x < 1, \\ 1 & \text{if } x \geq 1. \end{cases}$$

(c) Next we consider a discrete random variable X whose distribution function has the (finite or denumerable) set $\{x_j\}$ of discontinuity points with jumps p_j at the point x_j. The random variable X can be represented as the sum† of indicators,

$$X = \sum_j x_j I_{[X=x_j]}.$$

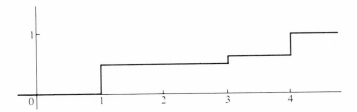

Figure 3.1. Example of a discrete distribution:
$F(x) = \tfrac{1}{2}\varepsilon(x-1) + \tfrac{1}{8}\varepsilon(x-3) + \tfrac{3}{8}\varepsilon(x-4)$.

The distribution function of X can be written in a form that is sometimes more convenient than (3.2.2), namely (see Figure 3.1)

$$F_X(x) = \sum_j p_j \, \varepsilon(x - x_j). \tag{3.3.4}$$

(d) We studied already (Theorem 2.3.1) the probability of the number of successes in n independent trials of an alternative with probability p of success. We define a random variable X_j which is equal to 1 or 0 depending whether the jth trial results in success or in failure. The number of successes in n trials

$$S_n = \sum_{j=1}^{n} X_j$$

---

† We write in the following $\sum_{j=1}^{n}$ if the summation runs from 1 to n, similarly $\sum_{j=1}^{\infty}$ if we sum from 1 to ∞. We write $\sum_j$, without specifying the range of summation if the sum is extended over a set which can be finite or denumerably infinite.

3.3 DISCRETE DISTRIBUTIONS

is then (Theorem 3.1.1) a random variable and its probabilities are (see Theorem 2.3.1)

$$P(S_n = k) = \binom{n}{k} p^k(1-p)^{n-k} \qquad (k = 0, 1, \ldots, n)$$

The distribution function of S_n is called the binomial distribution and is given by

$$F_{S_n}(x) = \sum_{k \leq x} \binom{n}{k} p^k (1-p)^{n-k} = \sum_{k=0}^{n} \binom{n}{k} p^k (1-p)^{n-k} \varepsilon(x-k). \tag{3.3.5}$$

(e) We discussed also sampling without replacement in the preceding section. Let X be the number of successes in k drawings, we see from (2.3.1) and (3.3.1) that

$$F_X(x) = \sum_{k=0}^{n} \frac{\binom{M}{k}\binom{N-M}{n-k}}{\binom{N}{n}} \varepsilon(x-k). \tag{3.3.6}$$

The distribution (3.3.6) is called the hypergeometric distribution. We note that the model which led to the hypergeometric distribution can be slightly modified: Instead of drawing n balls without replacement, one could draw n balls simultaneously.

(f) Next we consider distributions that have a denumerable set of discontinuity points.

Let the outcome space Ω be the sequence $\{\omega_n\}_{n=0}^{\infty}$ and let $\mathfrak{U}$ be the system of subsets of Ω. We select a number p such that $0 < p < 1$ and assign to the point ω_n of the outcome space the probability

$$p_n = (1-p)^n p \qquad (n = 0, 1, 2, \ldots, ad\ infinitum).$$

Since the p_n are all positive and $\sum_{n=0}^{\infty} p_n = 1$ this is a suitable assignment of probabilities.

Let $X = X(\omega) = n$ if $\omega = \omega_n$, this is a "labeling" random variable and $P(X = n) = p_n$. The distribution function of this random variable is

$$F_X(x) = \sum_{n=0}^{\infty} p_n \varepsilon(x - n) = p \sum_{n=0}^{\infty} (1 - p)^n \varepsilon(x - n). \qquad (3.3.7)$$

The probabilities p_n form a geometric series, the distribution (3.3.7) is therefore called the geometric distribution (sometimes also the Pascal distribution). The geometric distribution admits the following interpretation: Consider an infinite sequence of Bernoulli trials with p as probability of success and $1 - p$ as probability of failure. Then X is the number of failures preceding the first success in an infinite sequence of trials.

(g) As a second example we consider the same outcome space but assign to its points the probabilities

$$p_n = e^{-\lambda} \frac{\lambda^n}{n!} \qquad (n = 0, 1, \ldots, \textit{ad infinitum}; \quad \lambda > 0). \qquad (3.3.8)$$

These are also the probabilities of a labeling random variable, its distribution function is

$$F_X(x) = e^{-\lambda} \sum_{n=0}^{\infty} \frac{\lambda^n}{n!} \varepsilon(x - n). \qquad (3.3.9)$$

This is one of the most important discrete distributions, and is called the *Poisson distribution*. It is not only of great theoretical interest but also provides a useful model in a great variety of applications. We mention here only the radioactive disintegration of atoms, the counting of microorganisms in a suspension, problems of determining waiting times in many situations.

3.4 Examples of Absolutely Continuous Distributions

In this section, we introduce several absolutely continuous distributions by defining either their distribution functions or their frequency functions.

3.4 ABSOLUTELY CONTINUOUS DISTRIBUTIONS

(a) Let r be a positive and let a be a real number. The function

$$F(x) = \begin{cases} 0 & \text{if } x \leq a - r, \\ \dfrac{1}{2r}(x - a + r) & \text{if } a - r < x \leq a + r, \\ 1 & \text{if } x > a + r \end{cases} \quad (3.4.1)$$

is an absolutely continuous distribution function (see Figure 3.2); its frequency function is

$$p(x) = \begin{cases} \dfrac{1}{2r} & \text{if } |x - a| < r, \\ 0 & \text{if } |x - a| > r. \end{cases} \quad (3.4.2)$$

It is easily seen that $F(x) = \int_{-\infty}^{x} p(y)\, dy$. The distribution (3.4.1) is called the rectangular (or uniform) distribution.

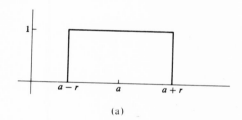

(a)

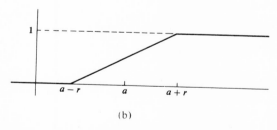

(b)

Figure 3.2. Rectangular (uniform) distributions. (a) Frequency function, (b) distribution function.

(b) Again let a and r be real numbers, $r > 0$; the function

$$p(x) = \begin{cases} \dfrac{1}{r^2}(r - |x - a|) & \text{if } |x - a| < r, \\ 0 & \text{if } |x - a| > r \end{cases} \quad (3.4.3)$$

is a frequency function; its distribution function is called the triangular (or Simpson's) distribution (see Figure 3.3).

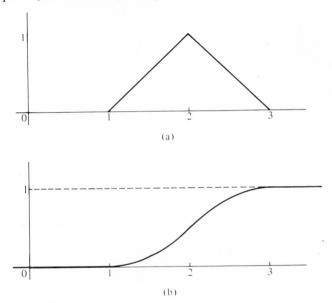

Figure 3.3 Triangular distributions (Simpson's distribution). (a) Frequency function, (b) distribution function. $a = 2$, $r = 1$.

The frequency functions of the rectangular and of the triangular distributions vanish outside a finite interval; we discuss next absolutely continuous distributions whose frequency functions are positive for all real values or all positive values.

(c) The function

$$p(x) = \tfrac{1}{2}e^{-|x|} \qquad (-\infty < x < \infty) \quad (3.4.4)$$

3.4 ABSOLUTELY CONTINUOUS DISTRIBUTIONS

is nonnegative and it is easily seen that $\int_{-\infty}^{\infty} p(x)\,dx = 1$ so that $p(x)$ is a frequency function. The corresponding distribution is

$$F(x) = \int_{-\infty}^{x} p(x)\,dx = \begin{cases} \tfrac{1}{2}e^{x} & \text{for } x < 0, \\ 1 - \tfrac{1}{2}e^{-x} & \text{for } x \geq 0. \end{cases} \quad (3.4.5)$$

This distribution is called the Laplace distribution.

(d) Let θ be an arbitrary positive constant. The function

$$F(x) = \begin{cases} 0 & \text{for } x < 0, \\ 1 - e^{-\theta x} & \text{for } x \geq 0 \end{cases} \quad (3.4.6)$$

is an absolutely continuous distribution function. Its frequency function is

$$p(x) = \begin{cases} 0 & \text{for } x < 0 \\ \theta e^{-\theta x} & \text{for } x > 0. \end{cases} \quad (3.4.7)$$

The distribution (3.4.6) is called the exponential (sometimes, negative exponential) distribution. It occurs in many applications, for instance in problems involving life testing of light bulbs and of similar items.

(e) Next we introduce one of the most important distributions, the Normal distribution.† It is known that

$$\int_{0}^{\infty} \exp\left(-\frac{y^{2}}{2}\right) dy = \sqrt{\frac{\pi}{2}}.$$

The function

$$\varphi(x) = \frac{1}{\sqrt{2\pi}} \exp\left(-\frac{x^{2}}{2}\right) \quad (-\infty < x < \infty) \quad (3.4.8)$$

is therefore a frequency function. The corresponding distribution function

$$\Phi(x) = \frac{1}{\sqrt{2\pi}} \int_{-\infty}^{x} \exp\left(-\frac{y^{2}}{2}\right) dy \quad (3.4.9)$$

---

† It is also called the Gaussian or the Gauss–Laplace distribution or—primarily in the French literature—the law of Moivre–Laplace or the second law of Laplace. The qualification "second law" is used to distinguish it from Laplace's first law which we called the Laplace distribution [see (c) (3.4.5)].

is called the standardized normal distribution (see Figure 3.4). The integral $\Phi(x)$ cannot be expressed in terms of elementary functions, however extensive tables of the normal distribution are available. (See Tables I and IA.)

The standardized normal distribution function $\Phi(x)$ satisfies the following inequality.

$$\frac{1}{\sqrt{2\pi}} \left(\frac{1}{x} - \frac{1}{x^3}\right) \exp\left(-\frac{x^2}{2}\right) < 1 - \Phi(x) < \frac{1}{\sqrt{2\pi}} \frac{1}{x} \exp\left(-\frac{x^2}{2}\right). \tag{3.4.10}$$

To prove the inequality we observe that

$$\frac{d}{dx}\left[x^{-1} \exp\left(-\frac{x^2}{2}\right)\right] = -\exp\left(-\frac{x^2}{2}\right)(1 + x^{-2})$$

so that

$$x^{-1} \exp\left(-\frac{x^2}{2}\right) = \int_x^\infty \exp\left(-\frac{y^2}{2}\right)(1 + y^{-2}) \, dy > \int_x^\infty \exp\left(-\frac{y^2}{2}\right) dy. \tag{3.4.11a}$$

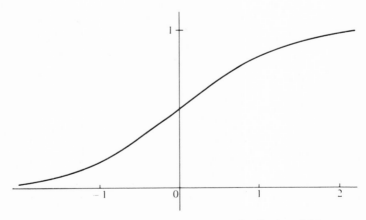

Figure 3.4. The standardized normal distribution $\Phi(x)$.

3.4 ABSOLUTELY CONTINUOUS DISTRIBUTIONS

Similarly we note that

$$\frac{d}{dx}\left[(x^{-1} - x^{-3})\exp\left(-\frac{x^2}{2}\right)\right] = -(1 - 3x^{-4})\exp\left(-\frac{x^2}{2}\right)$$

so that

$$(x^{-1} - x^{-3})\exp\left(-\frac{x^2}{2}\right) = \int_x^\infty \exp\left(-\frac{y^2}{2}\right)(1 - 3y^{-4})\,dy$$

$$< \int_x^\infty \exp\left(-\frac{y^2}{2}\right) dy. \qquad (3.4.11b)$$

Inequalities (3.4.10) follow immediately from the inequalities (3.4.11a) and (3.4.11b).

We conclude from (3.4.10) that

$$1 - \Phi(x) \sim \frac{1}{x\sqrt{2\pi}}\exp\left(-\frac{x^2}{2}\right)$$

for large x. (The symbol $\sim$ stands for approximately equal.)

The function

$$p(x) = \frac{1}{\sigma\sqrt{2\pi}}\exp\left[-\frac{1}{2\sigma^2}(x - \alpha)^2\right] \qquad (-\infty < x < \infty) \qquad (3.4.12)$$

with α real $\sigma^2 > 0$, is also a frequency function (see Figure 3.5). The corresponding distribution function

$$\int_{-\infty}^x p(y)\,dy = \frac{1}{\sigma\sqrt{2\pi}}\int_{-\infty}^x \exp\left[-\frac{1}{2\sigma^2}(y - \alpha)^2\right] dy = \Phi\left(\frac{x - \alpha}{\sigma}\right)$$

$$(3.4.13)$$

is called the normal distribution with parameters† α and σ.

---

† The parameter α is usually called the mean, while σ^2 is called the variance of the normal distribution (3.4.12). This terminology will be justified in Chapter 4, where we shall also discuss the significance of these parameters.

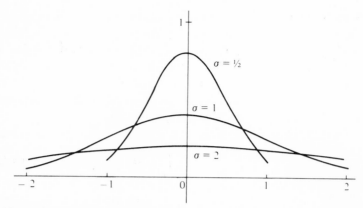

Figure 3.5. Normal frequency curves with mean 0, and $\sigma = \frac{1}{2}$, 1, and 2.

(*f*) It is known that

$$\int_0^\infty \frac{dx}{1+x^2} = \frac{\pi}{2},$$

the function

$$p(x) = \frac{1}{\pi(1+x^2)} \quad (-\infty < x < \infty) \qquad (3.4.14)$$

is therefore a frequency function, the corresponding distribution function is

$$F(x) = \frac{1}{2} + \frac{1}{\pi} \arctan x. \qquad (3.4.15)$$

This distribution is called the Cauchy distribution.

3.5 Multivariate Distributions

In many experiments, it is necessary to observe several quantities at the same time, since the object of the investigation cannot be described by a single number, but requires a number of separate measurements.

3.5 MULTIVARIATE DISTRIBUTIONS

Thus one might have to record the weight and height of every person belonging to a group. Educational measurements provide another example: Here one often has to record the scores of individual tests belonging to a battery of tests and also the age, sex, and years of formal education of each person tested.

If we wish to adapt the mathematical model to such situations we must study simultaneously several random variables.

Let $(\Omega, \mathfrak{U}, P)$ be a probability space, and let X and Y be two random variables defined on it. We can then make probability statements about the simultaneous behavior of X and Y. For instance, if we interpret X and Y as the rectangular coordinates of a point in the plane, we can introduce the probability that the point (X, Y) is located in a closed rectangle Δ with vertices (a_1, a_2), (b_1, a_2), (b_1, b_2), (a_1, b_2) since $[(X, Y) \in \Delta] = [a_1 \leq X \leq b_1] \cap [a_2 \leq Y \leq b_2] \in \mathfrak{U}$ is an event. For the sake of simplicity we write $P(a_1 \leq X \leq b_1, a_2 \leq Y \leq b_2)$ for its probability and use a similar notation if the rectangle is not closed.

It is again useful to define distribution functions for pairs of random variables. We write

$$F_{XY}(x, y) = P(X \leq x, Y \leq y)$$

and call $F_{XY}(x, y)$ the joint distribution of the random variables X and Y. Using the joint distribution we see easily that

$$P(a_1 < X \leq b_1, a_2 < Y \leq b_2)$$
$$= F_{XY}(b_1, b_2) - F_{XY}(b_1, a_2) - F_{XY}(a_1, b_2) + F_{XY}(a_1, a_2).$$
(3.5.1)

It is, of course, possible to consider simultaneously any number of random variables defined on the same probability space and to define their joint distribution. Let $X_1, \ldots, X_n$ be n random variables on $(\Omega, \mathfrak{U}, P)$, the function

$$F_{X_1, \ldots, X_n}(x_1, \ldots, x_n) = P(X_1 \leq x_1, \ldots, X_n \leq x_n)$$

is called the joint distribution of the random variables $X_1, \ldots, X_n$. We call joint distributions of $n > 1$ random variables multivariate distributions† (for $n = 2$ we also use the term bivariate distributions) to distinguish them from the distribution functions of single random variables to which we refer sometimes as univariate distributions. The properties of multivariate distributions are similar to those of univariate distributions, stated in Theorem 3.2.1. We present here, without proof, the result‡ for bivariate distributions.

Theorem 3.5.1

The joint distribution of the random variables X and Y has the following properties:

(i) $F_{XY}(b_1, b_2) - F_{XY}(b_1, a_2) - F_{XY}(a_1, b_2) + F_{XY}(a_1, a_2) \geq 0$, for all a_1, a_2, b_1, b_2.

(ii) $\lim_{x \to -\infty} F_{XY}(x, y) = \lim_{y \to -\infty} F_{XY}(x, y) = 0$;

$\lim_{x \to \infty} F_{XY}(x, y) = F_Y(y)$;

$\lim_{y \to \infty} F_{XY}(x, y) = F_X(x)$;

$\lim_{x \to \infty} \lim_{y \to \infty} F_{XY}(x, y) = 1$.

(iii) $F_{XY}(x, y)$ *is right continuous in each of the variables x and y.*

Theorem 3.5.1 can be extended to the joint distribution of n random variables. Condition (i), which is a consequence of Axiom (IIA), then becomes more complicated.

We shall again consider two disjoint classes of multivariate distributions, namely discrete and absolutely continuous distributions. These two classes do not exhaust all possibilities but the most important distribution functions belong to one of these classes.

The joint distribution of the random variables X and Y is discrete if each of the random variables has a discrete distribution.

---

† Multivariate distributions are sometimes called n-dimensional distributions.
‡ See [1, p. 78].

3.5 MULTIVARIATE DISTRIBUTIONS

Let $\{x_i\}$ and $\{y_j\}$ be the possible values† of X and Y, respectively; we write $p_{ij} = P(X = x_i, Y = y_j)$, where $p_{ij} \geq 0$ for all i and j and $\sum_i \sum_j p_{ij} = 1$. The individual probabilities of X and Y are called marginal probabilities, and are given by $P(X = x_i) = \sum_j p_{ij}$ and $P(Y = y_j) = \sum_i p_{ij}$. We write $p_i. = \sum_j p_{ij}$ and $p._j = \sum_i p_{ij}$.

More generally we say that the joint distribution of the n random variables $X_1, \ldots, X_n$ is discrete if each of the X_j has a discrete distribution.

Let $\{x_{il}\}$ ($l = 1, 2, \ldots, n$) be the countable set of possible values of the random variable X_l, we write

$$p_{i_1 i_2 \cdots i_n} = P(X_1 = x_{i_1,1}, X_2 = x_{i_2,2}, \ldots, X_n = x_{i_n,n})$$

and can define k-dimensional marginal distributions by summations.

We introduce another important concept and consider, for the sake of simplicity, only the bivariate case. Let X and Y be two discrete random variables with possible values $\{x_i\}$ and $\{y_j\}$ and let $p_{ij} = P(X = x_i, Y = y_j)$. Suppose that the marginal probability $p._j = \sum_i p_{ij} > 0$. We define the conditional probability that $X = x_i$ if it is known that $Y = y_j$ by

$$P(X = x_i | Y = y_j) = \frac{p_{ij}}{p._j}. \tag{3.5.2}$$

This is in agreement with (2.2.1), with $A = [X = x_i]$ and $B = [Y = y_j]$. The conditional probability $P(Y = y_j | X = x_i)$ as well as conditional probabilities in the case of n random variables are defined in a similar manner.

The joint distribution $F_{X_1 \cdots X_n}(x_1, \ldots, x_n)$ of the n random variables $X_1, \ldots, X_n$ is said to be absolutely continuous if there exists a function

---

† The sets $\{x_i\}$ and $\{y_j\}$ are both countable.

$p_{X_1\cdots X_n}(x_1,\ldots,x_n)$ such that

$$F_{X_1\cdots X_n}(x_1,\ldots,x_n) = \int_{-\infty}^{x_1}\cdots\int_{-\infty}^{x_n} p_{X_1\cdots X_n}(y_1,\ldots,y_n)\,dy_1\cdots dy_n.$$
(3.5.3)

The function $p_{X_1\cdots X_n}$ is again called the frequency function of the distribution $F_{X_1\cdots X_n}$ and the relation

$$p_{X_1\cdots X_n}(x_1,\ldots,x_n) = \frac{\partial^n}{\partial x_1\cdots\partial x_n} F_{X_1\cdots X_n}(x_1,\ldots,x_n) \quad (3.5.4)$$

holds.

Multivariate frequency functions satisfy the following conditions:

$$p(x_1,\ldots,x_n) \geq 0,$$

$$\int_{-\infty}^{\infty}\cdots\int_{-\infty}^{\infty} p(x_1,\ldots,x_n)\,dx_1\cdots dx_n = 1.$$

These relations correspond to (3.2.3a) and (3.2.3b) and also assure the existence of random variables $X_1,\ldots,X_n$ having an absolutely continuous distribution whose frequency function is $p(x_1,\ldots,x_n)$.

We consider a bivariate absolutely continuous distribution

$$F_{X_1 X_2}(x_1, x_2) = \int_{-\infty}^{x_1}\int_{-\infty}^{x_2} p_{X_1 X_2}(y_1, y_2)\,dy_1\,dy_2.$$

Since the marginal distribution of X_1 is

$$F_{X_1}(x_1) = \lim_{x_2\to\infty} F_{X_1 X_2}(x_1, x_2) = \int_{-\infty}^{x_1}\int_{-\infty}^{\infty} p_{X_1 X_2}(y_1, y_2)\,dy_2\,dy_1$$

we see that the marginal distribution of X_1 is absolutely continuous and that

$$p_{X_1}(x_1) = \int_{-\infty}^{\infty} p_{X_1 X_2}(x_1, y)\,dy. \qquad (3.5.5a)$$

3.5 MULTIVARIATE DISTRIBUTIONS

Similarly

$$p_{X_2}(x_2) = \int_{-\infty}^{\infty} p_{X_1 X_2}(y, x_2)\, dy. \qquad (3.5.5b)$$

In the n-dimensional case the marginal frequencies are also obtained by integrations.

We next define conditional probabilities for absolutely continuous random variables. Let $p_{X_1 X_2}(x_1, x_2)$ be the bivariate frequency function of the random variables X_1 and X_2 and suppose that x_2 is a value for which

$$p_{X_2}(x_2) > 0.$$

We define the conditional frequency function of X_1 given that $X_2 = x_2$ as

$$p_{X_1}(x_1 \mid X_2 = x_2) = \frac{p_{X_1 X_2}(x_1, x_2)}{p_{X_2}(x_2)} \qquad (3.5.6a)$$

and similarly—if $p_{X_1}(x_1) > 0$—the conditional frequency

$$p_{X_2}(x_2 \mid X_1 = x_1) = \frac{p_{X_1 X_2}(x_1, x_2)}{p_{X_1}(x_1)}. \qquad (3.5.6b)$$

One can easily show that the conditional frequencies satisfy (3.2.3a) and (3.2.3b) so that they are indeed frequency functions.

We can also introduce conditional distribution functions for absolutely continuous random variables as integrals of their conditional frequency functions. We obtain these from (3.5.6a) and (3.5.6b), respectively.

$$F_{X_1}(x \mid X_2 = x_2) = \int_{-\infty}^{x} p_{X_1}(y \mid X_2 = x_2)\, dy$$

$$= \frac{1}{p_{X_2}(x_2)} \int_{-\infty}^{x} p_{X_1 X_2}(y, x_2)\, dy$$

and similarly

$$F_{X_2}(x \mid X_1 = x_1) = \frac{1}{p_{X_1}(x_1)} \int_{-\infty}^{x} p_{X_1 X_2}(x_1, y)\, dy.$$

If X_1 and X_2 are independent random variables then

$$p_{X_1, X_2}(x_1, x_2) = p_{X_1}(x_1) p_{X_2}(x_2),$$

and we see that

$$F_{X_1}(x \mid X_2 = x_2) = F_{X_1}(x),$$

and similarly

$$F_{X_2}(x \mid X_1 = x_1) = F_{X_2}(x).$$

As examples we introduce two important multivariate distributions. The first is discrete and is a generalization of the binomial distribution, the second is absolutely continuous and is the multivariate analog of the normal distribution.

(a) We consider n independent trials of an experiment that has m exhaustive and mutually exclusive outcomes $A_1, A_2, \ldots, A_m$. Let $p_j = P(A_j)$, $j = 1, 2, \ldots, m$, be the probability of the outcome A_j in a single trial. Since the A_j are exhaustive and mutually exclusive we have

$$\sum_{j=1}^{m} p_j = 1.$$

There are m^n possible outcomes of this experiment. These can be described by m-tuples $(k_1, \ldots, k_m)$ of nonnegative integers, indicating that outcome A_j was obtained k_j times in the n trials. Clearly

$$\sum_{j=1}^{m} k_j = n.$$

We wish to determine the probability $P_n(k_1, \ldots, k_m)$ that the n trials result in observing k_1 times the outcome A_1, k_2 times the outcome $A_2, \ldots, k_m$ times the outcome A_m. This probability is independent of the order in which the outcomes occur. In order to find $P_n(k_1, \ldots, k_m)$ we first determine the probability that A_1 occurs k_1 times, A_2 occurs k_2 times, $\ldots$, A_m occurs k_m times in a completely specified order. This probability is given by

$$p_1^{k_1} p_2^{k_2} \cdots p_m^{k_m}.$$

3.5 MULTIVARIATE DISTRIBUTIONS

The probability $P_n(k_1, \ldots, k_m)$ is obtained by multiplying $p_1^{k_1} p_2^{k_2} \cdots p_m^{k_m}$ with the number of ways in which the order can be specified, if it is known that A_j occurs k_j times in n trials. Denote this number by $\binom{n}{k_1 \cdots k_m}$ then

$$\binom{n}{k_1 \cdots k_m} = \binom{n}{k_1}\binom{n-k_1}{k_2} \cdots \binom{n - k_1 - k_2 - \cdots - k_{m-1}}{k_m}.$$

We express the binomial coefficients in terms of factorials and see that

$$\binom{n}{k_1 \cdots k_m}$$

$$= \frac{n!}{k_1!(n-k_1)!} \cdot \frac{(n-k_1)!}{k_2!(n-k_1-k_2)!} \cdots \frac{(n-k_1-\cdots-k_{m-1})!}{k_m!}$$

or

$$\binom{n}{k_1 \cdots k_m} = \frac{n!}{k_1! \cdots k_m!}.$$

Therefore

$$P_n(k_1, \ldots, k_m) = \frac{n!}{k_1! \cdots k_m!} p_1^{k_1} \cdots p_m^{k_m}. \quad (3.5.7)$$

The probabilities $P_n(k_1, \ldots, k_m)$ are called multinomial probabilities since they are the coefficients of the multinomial expansion of

$$(p_1 + p_2 + \cdots + p_m)^n.$$

We define random variables $X_1, \ldots, X_m$ such that X_j denotes the number of times A_j occurs in n trials. Then

$$P(X_1 = k_1, X_2 = k_2, \ldots, X_m = k_m) = P_n(k_1, k_2, \ldots, k_m) \quad (3.5.8)$$

and the joint distribution of $X_1, X_2, \ldots, X_m$, given by (3.5.7), is called the multinomial distribution. For $m = 2$, it reduces to the binomial distribution.

(b) Let $Q = Q(u, v) = a_{11}u^2 - 2a_{12}uv + a_{22}v^2$ be a positive-definite quadratic form.† Then there exists a constant C such that $C\mathcal{T} = 1$ where

$$\mathcal{T} = \int_{-\infty}^{\infty}\int_{-\infty}^{\infty} \exp\{-\tfrac{1}{2}[a_{11}(x-m_1)^2 - 2a_{12}(x-m_1)(y-m_2)$$
$$+ a_{22}(y-m_2)^2]\}\, dx\, dy. \qquad (3.5.9)$$

To evaluate the integral $\mathcal{T}$ we put

$$u = x - m_1, \qquad v = y - m_2 \qquad (3.5.10)$$

so that

$$\mathcal{T} = \int_{-\infty}^{\infty}\int_{-\infty}^{\infty} \exp[-\tfrac{1}{2}Q(u,v)]\, du\, dv.$$

Since $Q(u,v) = \dfrac{A}{a_{22}} u^2 + a_{22}\left(v - \dfrac{a_{12}}{a_{22}} u\right)^2$, we obtain for the integral

$$\int_{-\infty}^{\infty} \exp\left(-\tfrac{1}{2}Q\right) dv = \exp\left[-\dfrac{A}{2a_{22}} u^2\right] \int_{-\infty}^{\infty} \exp\left[-\dfrac{a_{22}}{2}\left(v - \dfrac{a_{12}}{a_{22}} u\right)^2\right] dv$$

$$= \sqrt{\dfrac{2\pi}{a_{22}}} \exp\left[-\dfrac{A}{2a_{22}} u^2\right]. \qquad (3.5.11)$$

A simple computation shows then that

$$\int_{-\infty}^{\infty}\int_{-\infty}^{\infty} \exp\left[-\tfrac{1}{2}Q(u,v)\right] du\, dv = \dfrac{2\pi}{\sqrt{A}}$$

so that $C = \dfrac{\sqrt{A}}{2\pi}$, where $A = a_{11}a_{22} - a_{12}^2$. The function

$$p(x,y) = \dfrac{\sqrt{A}}{2\pi} \exp\left\{-\tfrac{1}{2}[a_{11}(x-m_1)^2 - 2a_{12}(x-m_1)(y-m_2) + a_{22}(y-m_2)^2]\right\} \qquad (3.5.12)$$

---

† This means that $A = a_{11}a_{22} - a_{12}^2 > 0$, Q vanishes then only if $u = v = 0$.

3.5 MULTIVARIATE DISTRIBUTIONS

is a bivariate frequency function. The corresponding distribution function is called the bivariate normal distribution.

We compute next the marginal frequency functions of this distribution. Let X and Y be two random variables that have a bivariate normal distribution. To simplify the notation we write $p_1(x)$ [respectively $p_2(y)$] for the marginal frequency of X (respectively Y). We see from (3.5.5a) that

$$p_1(x) = \int_{-\infty}^{\infty} p(x, y) \, dy = \frac{\sqrt{A}}{2\pi} \int_{-\infty}^{\infty} \exp\left[-\frac{1}{2} Q(x - m_1, y - m_2)\right] dy.$$

We make again the substitution $v = y - m_2$ and use (3.5.11) and see that

$$p_1(x) = \frac{1}{\sqrt{2\pi}} \sqrt{\frac{A}{a_{22}}} \exp\left[-\frac{A}{2a_{22}} (x - m_1)^2\right]. \qquad (3.5.13a)$$

The marginal distribution $p_1(x)$ is therefore normal with mean m_1 and variance $\frac{a_{22}}{A}$. Similarly, one obtains the marginal distribution of Y. This is again normal with mean m_2 and variance $\frac{a_{11}}{A}$ and is given by

$$p_2(y) = \frac{1}{\sqrt{2\pi}} \sqrt{\frac{A}{a_{11}}} \exp\left[-\frac{A}{2a_{11}} (y - m_2)^2\right]. \qquad (3.5.13b)$$

We compute also the conditional frequency functions and use for the sake of simplicity the notation $p(x|y)$ and $p(y|x)$ instead of $p_X(x|Y = y)$ and $p_Y(y|X = x)$. We see from (3.5.13a) and (3.5.13b) that the marginal distributions are always positive so that $p(x|y)$ and $p(y|x)$ are defined for all values of x and y. We have

$$p(y|x) = \frac{p(x, y)}{p_1(x)} \qquad (3.5.14)$$

where $p(x, y)$ is given by (3.5.12) while $p_1(x)$ is given by (3.5.13a).

Let σ_1^2 (respectively σ_2^2) be the variance of X (respectively Y). We see from (3.5.13a) and (3.5.13b) that

$$\sigma_1^2 = \frac{a_{22}}{A}, \qquad \sigma_2^2 = \frac{a_{11}}{A}, \qquad (3.5.15a)$$

where $A = a_{11}a_{22} - a_{12}^2$. It is convenient to change the parameters in (3.5.12) by introducing σ_1, σ_2 and a third parameter ρ instead of a_{11}, a_{12}, and a_{22}. We define

$$\rho^2 = \frac{a_{12}^2}{a_{11}a_{22}} \qquad (3.5.15b)$$

then $1 - \rho^2 = \dfrac{A}{a_{11}a_{22}}$ and therefore

$$a_{11} = \frac{1}{(1-\rho^2)\sigma_1^2}, \qquad a_{22} = \frac{1}{(1-\rho^2)\sigma_2^2}. \qquad (3.5.16a)$$

It follows from (3.5.15b) and (3.5.16a) that

$$a_{12} = \frac{\rho}{(1-\rho^2)\sigma_1\sigma_2}. \qquad (3.5.16b)$$

Here ρ may have any sign. We also see that

$$A = \frac{1}{(1-\rho^2)\sigma_1^2\sigma_2^2}. \qquad (3.5.16c)$$

We substitute (3.5.16a), (3.5.16b), and (3.5.16c) into (3.5.12) and obtain

$$p(x, y) = \frac{1}{2\pi\sigma_1\sigma_2\sqrt{1-\rho^2}}$$

$$\times \exp\left\{-\frac{1}{2(1-\rho^2)}\left[\frac{(x-m_1)^2}{\sigma_1^2} - \frac{2\rho(x-m_1)(y-m_2)}{\sigma_1\sigma_2} + \frac{(y-m_2)^2}{\sigma_2^2}\right]\right\}.$$

$$(3.5.17)$$

3.5 MULTIVARIATE DISTRIBUTIONS

We see from (3.5.13a) and (3.5.15a) that

$$p_1(x) = \frac{1}{\sigma_1\sqrt{2\pi}} \exp\left[-\frac{1}{2\sigma_1^2}(x - m_1)^2\right]. \quad (3.5.18)$$

An elementary computation, using (3.5.14), (3.5.17), and (3.5.18), yields finally

$$p(y \mid x) = \frac{1}{\sigma_2\sqrt{2\pi(1 - \rho^2)}}$$

$$\times \exp\left\{-\frac{1}{2(1-\rho^2)\sigma_2^2}\left[y - m_2 - \frac{\rho\sigma_2}{\sigma_1}(x - m_1)\right]^2\right\}. \quad (3.5.19a)$$

The conditional distribution of Y, given X, is therefore normal with mean $m_2 + \frac{\rho\sigma_2}{\sigma_1}(x - m_1)$ and variance $(1 - \rho^2)\sigma_2^2$.

The conditional distribution of X, given Y, is obtained in the same way; one has

$$p(x \mid y) = \frac{1}{\sigma_1\sqrt{2\pi(1 - \rho^2)}}$$

$$\times \exp\left\{-\frac{1}{2(1-\rho^2)\sigma_1^2}\left[x - m_1 - \frac{\rho\sigma_1}{\sigma_2}(y - m_2)\right]^2\right\}. \quad (3.5.19b)$$

This is again a normal distribution with mean $m_1 + \frac{\rho\sigma_1}{\sigma_2}(y - m_2)$ and variance $(1 - \rho^2)\sigma_1^2$. We know already that σ_1^2 (respectively σ_2^2) is the variance of X (respectively Y) and a similar interpretation of the parameter ρ will be given in Section 4.4.

The preceding discussion can be extended to the case of n variables. Let $Q(x_1, \ldots, x_n) = \sum_{i=1}^{n} \sum_{j=1}^{n} a_{ij} x_i x_j$ be a positive-definite quadratic form.†

---

† This means that the determinant of the coefficients $A = |a_{ij}| > 0$. Then $Q(x_1, \ldots, x_n) = 0$ only if all the $x_j = 0, j = 1, \ldots, n$.

It is no restriction to assume that $a_{ij} = a_{ji}$. It can be shown that the function

$$p(x_1, \ldots, x_n) = \sqrt{A}(2\pi)^{-n/2} \exp[-\tfrac{1}{2} Q(x_1, \ldots, x_n)]$$

is a frequency function. The corresponding distribution function is called the multivariate normal distribution.

Remark

The marginal distributions of the bivariate (also of the n-variate) normal distribution are normal distributions. However, the converse is not true. There exist nonnormal multivariate distributions which have normal marginals.

We introduce a very important concept of probability theory.

The n random variables $X_1, \ldots, X_n$ are said to be independent if the n events $[a_1 \leq X_1 \leq b_1], \ldots, [a_n \leq X_n \leq b_n]$ are independent for all $a_1, b_1, \ldots, a_n, b_n$. Then

$$P(X_1 \leq x_1, \ldots, X_n \leq x_n) = \prod_{j=1}^{n} P(X_j \leq x_j)$$

or, in terms of distribution functions

$$F_{X_1 \cdots X_n}(x_1, \ldots, x_n) = \prod_{j=1}^{n} F_{X_j}(x_j). \qquad (3.5.20)$$

The joint distribution function of n independent random variables is therefore the product of their one-dimensional marginal distributions.

If the joint distribution of the $X_1, \ldots, X_n$ is discrete, (3.5.20) is equivalent to

$$P(X_1 = x_1, \ldots, X_n = x_n) = \prod_{j=1}^{n} P(X_j = x_j). \qquad (3.5.21)$$

If the joint distribution of the $X_1, \ldots, X_n$ is absolutely continuous, one can differentiate (3.5.20) once with respect to each variable x_j and see that the independence of the $X_1 \cdots X_n$ means, in terms of frequency functions, that the relation

$$p_{X_1 \cdots X_n}(x_1, \ldots, x_n) = \prod_{j=1}^{n} p_{X_j}(x_j) \qquad (3.5.22)$$

holds for all $x_1, \ldots, x_n$.

3.6 Problems

1. Let X be a random variable. Find the distribution function of the random variable $Y = X + a$ where a is a constant.
2. Let X be an absolutely continuous random variable and let $a \neq 0$ and b be two constants. Show that $Y = aX + b$ is also an absolutely continuous random variable and find the frequency function of Y.
3. Let X be a random variable. Show that $Y = |X|$ is also a random variable and find its distribution.
4. Let X be a random variable which assumes only positive values. Show that the square root of X (taken with the positive sign) is also a random variable. Determine the distribution function of $Y = \sqrt{X}$. Suppose further that X is absolutely continuous and determine the frequency function of Y.
5. Suppose that X has an exponential distribution. Find the distribution function and the frequency function of $Y = X^2$.
6. Consider a probability space $(\Omega, \mathfrak{U}, P)$. Let $A \in \mathfrak{U}$ and let I_A be the indicator function of A. Determine the distribution function of I_A.
7. Determine the constant k so that $p(x) = k(1 + x)^{-2}$ if $0 < a < x < b$ but $p(x) = 0$ otherwise, becomes a frequency function.
8. Let $F_{XY}(x, y)$ be a bivariate distribution and prove that

$$\lim_{x \to \infty} F_{XY}(x, y) = F_Y(y),$$

as stated in Theorem 3.5.1.

9. Show that the function

$$p(x, y) = \begin{cases} 4xy & \text{if } 0 < x < 1 \text{ and } 0 < y < 1, \\ 0 & \text{otherwise} \end{cases}$$

is a frequency function. Find the marginal frequency functions and the marginal distribution functions.

10. Let the random variables X and Y have the bivariate frequency function

$$p_{XY}(x, y) = \begin{cases} k(x^2 + y^2) & \text{if } 0 \le x < 1 \text{ and } 0 \le y < 1, \\ 0 & \text{otherwise.} \end{cases}$$

Determine the constant k.

11. Two continuous random variables X and Y have the joint frequency distribution

$$p_{XY}(x, y) = \begin{cases} xe^{-(x+y)} & \text{if } x \ge 0 \text{ and } y \ge 0, \\ 0 & \text{otherwise.} \end{cases}$$

Are X and Y independent? Give reasons for your answer.

12. Let the frequency function of X and Y be

$$p_{XY}(x, y) = \begin{cases} 2 & \text{if } 0 < x < y \text{ and } 0 < y < 1. \\ 0 & \text{otherwise.} \end{cases}$$

Are X and Y independent? Give reasons for your answer.

13. A straight line of length $2a$ is randomly divided† into two parts. What is the probability that the area of a rectangle formed from these two parts is less then $\frac{1}{2}a^2$?

14. Two points X and Y are selected independently at random† on the segment $(0, 1)$. What is the probability that the segments $0X$, YX, and $Y1$ can form a triangle?

15. Meteorological observations taken in a certain location during the month of July indicate that the probability of a thunderstorm is 0.3, while the probability of hail is 0.25. The probability of hail during a thunderstorm is 0.4. Find the probability
    (a) that there will be hail but no thunderstorm,
    (b) that there is hail, given that there was no thunderstorm, in this locality in July.

16. Let $G(x)$ be a continuous distribution function. Prove that $[G(x)]^3$ is also a distribution function.

---

† Randomly divided (or selected at random) means that the point dividing the line segment is uniformly distributed on the segment.

17. Two balls are drawn from an urn containing three white, five red, and seven green balls. What is the probability that the first ball is white and the second green if the drawing is carried out
    (a) without replacement,
    (b) with replacement?
18. The random variable X has the frequency function
$$F'(x) = \begin{cases} e^{-x} & \text{for } x \geq 0, \\ 0 & \text{for } x < 0. \end{cases}$$
Let $X = \log Y$ and find the frequency function of the random variable Y.
19. An urn contains 18 white and two black balls. Suppose that m balls ($m \leq 18$) are drawn at random *without replacement*.
    (a) What is the probability that at least one of the black balls is drawn?
    (b) What is the smallest value of m for which this probability will exceed $\frac{1}{2}$?
20. An urn contains 18 white and two black balls. Suppose that m balls are drawn at random *with replacement*.
    (a) What is the probability that at least one of the black balls is drawn?
    (b) What is the smallest value of m for which this probability will exceed $\frac{1}{2}$? (*Note*: $\log 2 = 0.30103$, $\log 3 = 0.47712$.)
21. A bar is rotated about its midpoint in such a way that the probability of a given end stopping at a given angle from its starting point is equal for all angles from $-\frac{1}{2}\pi$ to $\frac{1}{2}\pi$. Find the probability density function of the tangent of the angle.

References

1. H. Cramér, "Mathematical Methods of Statistics." Princeton Univ. Press, Princeton, New Jersey, 1946.
2. H. G. Tucker, "A Graduate Course in Probability." Academic Press, New York, 1967.

4 | Typical Values

4.1 The Mathematical Expectation of a Random Variable

The behavior of a random variable is characterized completely by its distribution function. However, it is often preferable to have a summary description by a few typical values, even if incomplete. In the present chapter we define a number of useful typical values of random variables and study their properties.

In this discussion, we shall treat discrete and absolutely continuous random variables separately. This procedure is necessary because of our intention to avoid, so far as possible, the use of advanced mathematical techniques. With the aid of these techniques it would be possible to give a unified treatment that would include not only the discrete and the absolutely continuous case, but would cover all distribution functions.

Let $(\Omega, \mathfrak{U}, P)$ be a probability space and consider a discrete random variable X defined on it. Let $\{x_j\}$ be the countable set of discontinuity points of the distribution function $F_X(x)$ of X and let $p_j = P(X = x_j)$. If the sum

$$\sum_j |x_j| p_j$$

converges then we say that the mathematical expectation $\mathscr{E}(X)$ (or expected value or mean) of the random variable X exists and write

$$\mathscr{E}(X) = \sum_j x_j p_j. \qquad (4.1.1)$$

Remark

We require the absolute convergence of the series (4.1.1) for the existence of the expectation. If the series is only conditionally convergent then the expectation of X does not exist.

Example 4.1.1

Let $X = I_A$ be the indicator random variable of a set $A \in \mathfrak{U}$. Then $\mathscr{E}(X) = 0 \cdot P(A^c) + 1 \cdot P(A) = P(A)$.

Example 4.1.2

Let X be the number of points obtained in a single toss of a die. Then $\mathscr{E}(X) = \sum_{j=1}^{6} j \tfrac{1}{6} = \tfrac{7}{2}$.

Example 4.1.3

Let X be a random variable that has a geometric distribution, that is, let $P(X = j) = pq^{j-1}$, where $q = 1 - p$ and $j = 1, 2, \ldots$, *ad infinitum*.

4.1 EXPECTATION OF A RANDOM VARIABLE

The possible values are all positive so that the expectation of X exists if the series (4.1.1) converges. We have

$$\mathscr{E}(X) = p \sum_{j=1}^{\infty} jq^{j-1} = p \frac{d}{dq} \left(\sum_{j=1}^{\infty} q^j \right) = p \frac{d}{dq} \left[\frac{q}{1-q} \right] = \frac{p}{(1-q)^2}$$

so that $\mathscr{E}(X) = \dfrac{1}{p}$.

Example 4.1.4

Let X be a random variable that assumes the values $x_j = (-1)^{j+1} \dfrac{3^j}{j}$ with probabilities $p_j = \dfrac{2}{3^j}$ ($j = 1, 2, \ldots$, *ad infinitum*). The sum

$$\sum_{j=1}^{\infty} |x_j| p_j = \sum_{j=1}^{\infty} \frac{3^j}{j} \frac{2}{3^j} = \sum_{j=1}^{\infty} \frac{2}{j}$$

diverges so that the expectation of X does not exist although the series

$$\sum_{j=1}^{\infty} \frac{(-1)^{j+1} 3^j}{j} \frac{2}{3^j} = 2 \sum_{j=1}^{\infty} (-1)^{j-1} \frac{1}{j}$$

is convergent.

The distribution functions of the random variables in Examples 4.1.3 and 4.1.4 have denumerable sets of possible values.

We consider next an absolutely continuous random variable X defined on our probability space. Let $p_X(x)$ be the frequency function of X. If the integral

$$\int_{-\infty}^{\infty} |x| p_X(x) \, dx$$

converges, we say that the expectation of X (respectively of the distribution function of X) exists and define the expectation $\mathscr{E}(X)$ by

$$\mathscr{E}(X) = \int_{-\infty}^{\infty} x p_X(x) \, dx. \tag{4.1.2}$$

Formula (4.1.2) is analogous to formula (4.1.1). We emphasize that the existence of the expectation requires the absolute convergence of the defining sum or integral.

Example 4.1.5

Let X be a random variable that has a normal distribution with parameters α and σ. It is easily seen that the integral

$$\frac{1}{\sigma\sqrt{2\pi}} \int_{-\infty}^{\infty} |x| \exp\left[-\frac{1}{2\sigma^2}(x-\alpha)^2\right] dx$$

$$= \frac{2}{\sigma\sqrt{2\pi}} \int_{0}^{\infty} x \exp\left[-\frac{1}{2\sigma^2}(x-\alpha)^2\right] dx$$

exists. The expected value of X is

$$\mathscr{E}(X) = \frac{1}{\sigma\sqrt{2\pi}} \int_{-\infty}^{\infty} x \exp\left[-\frac{1}{2\sigma^2}(x-\alpha)^2\right] dx$$

$$= \frac{\sigma}{\sqrt{2\pi}} \int_{-\infty}^{\infty} y \exp\left(-\frac{y^2}{2}\right) dy + \alpha \frac{1}{\sqrt{2\pi}} \int_{-\infty}^{\infty} \exp\left(-\frac{y^2}{2}\right) dy.$$

The first integral is zero while

$$\frac{1}{\sqrt{2\pi}} \int_{-\infty}^{\infty} \exp\left(-\frac{y^2}{2}\right) dy = 1 \qquad (4.1.3)$$

so that

$$\mathscr{E}(X) = \alpha. \qquad (4.1.4)$$

The parameter α is called the mean of the normal distribution.

Example 4.1.6

Suppose that the random variable X has a Cauchy distribution. It is easily seen that the integral $\int_{-\infty}^{\infty} \frac{|x|\, dx}{1+x^2}$ is divergent so that the Cauchy distribution has no expected value.

4.1 EXPECTATION OF A RANDOM VARIABLE

Expectations for multivariate distributions can be defined in a similar way. Let X and Y be two discrete random variables and suppose that X and Y have the possible values $\{x_i\}$ and $\{y_j\}$, respectively, and let $p_{ij} = P(X = x_i, Y = y_j)$. The expectations of X and Y are then defined by

$$\mathscr{E}(X) = \sum_i \sum_j x_i p_{ij} = \sum_i x_i p_{i\cdot},$$
$$\mathscr{E}(Y) = \sum_i \sum_j y_j p_{ij} = \sum_j y_j p_{\cdot j},$$
(4.1.5)

provided that the double sums are absolutely convergent.

If X and Y are absolutely continuous with frequency function $p_{XY}(x, y)$ then we define

$$\mathscr{E}(X) = \int_{-\infty}^{\infty} \int_{-\infty}^{\infty} x p_{XY}(x, y) \, dx \, dy = \int_{-\infty}^{\infty} x p_X(x) \, dx,$$
$$\mathscr{E}(Y) = \int_{-\infty}^{\infty} \int_{-\infty}^{\infty} y p_{XY}(x, y) \, dx \, dy = \int_{-\infty}^{\infty} y p_Y(y) \, dy,$$
(4.1.6)

provided that the integrals are absolutely convergent. Here $p_X(x)$ and $p_Y(y)$ are the marginal distributions of X and Y, respectively.

We now derive a few important properties of expectations.

A random variable X is said to be bounded if its distribution function has no point of increase outside a finite interval. If X is a discrete and bounded random variable with possible values $\{x_j\}$ then there exists a constant M such that $|x_j| \le M$ for all j. If X is absolutely continuous and bounded, there exists a constant M such that $p_X(x) = 0$ for $|x| > M$, hence $\int_{-M}^{M} p_X(x) \, dx = 1$. The constant M is called an upper bound of the random variable X.

Theorem 4.1.1

The expectation of a bounded random variable always exists.

In case X is discrete we have

$$\sum_j |x_j| p_j \le M \sum_j p_j = M,$$

in case X is absolutely continuous we have

$$\int_{-\infty}^{\infty} |x| p_X(x)\, dx = \int_{-M}^{M} |x| p_X(x)\, dx \leq M \int_{-M}^{M} p_X(x)\, dx = M$$

so that the statement is proven.

Theorem 4.1.2

Let C be an arbitrary constant. Then $\mathscr{E}(C) = C$.

This follows from the fact that C can be regarded as a (degenerate) random variable that assumes its only possible value C with probability 1.

Theorem 4.1.3

Let C be an arbitrary constant and let X be a random variable whose expectation exists, then $\mathscr{E}(CX) = C\mathscr{E}(X)$.

This follows immediately from (4.1.1) and (4.1.2).

4.2 Expectations of Functions of Random Variables

We saw (Theorems 3.1.2 and 3.1.3) that some functions of random variables are again random variables. However, not *all* functions of random variables are again random variables. Let $g(x)$ be a real-valued function that is defined for all real x, and let $X = X(\omega)$ be a random variable defined on the probability space $(\Omega, \mathfrak{U}, P)$. The statement that $g(X) = g[X(\omega)]$ is a random variable means that

$$\{\omega : g[X(\omega)] \leq x\} \in \mathfrak{U} \tag{4.2.1}$$

for all real x and any random variable X. In what follows, we shall

4.2 EXPECTATIONS OF FUNCTIONS

denote the class of all functions $g(x)$ that satisfy (4.2.1) by $\mathfrak{M}$. Condition (4.2.1) is certainly a restriction imposed on $g(x)$. Nevertheless, the class $\mathfrak{M}$ is still a very large class. $\mathfrak{M}$ contains, for example, all continuous functions, all piecewise continuous functions and many discontinuous functions. In this text we shall not treat the problem of describing the class $\mathfrak{M}$ but shall, whenever necessary, assume that a function $g(x)$ belongs to $\mathfrak{M}$. The class $\mathfrak{M}$ is so wide that every function that occurs in practical applications belongs to it.

The assumption that a function $g(x)$ belongs to $\mathfrak{M}$ does not assure the existence of the expectation of $g(X)$. The present section deals with the problem of determining the expectation of $g(X)$, if it exists.

Theorem 4.2.1

Let X be a discrete random variable with a countable set of possible values $\{x_j\}$, and let $p_j = P(X = x_j)$ and let $g(x) \in \mathfrak{M}$. The expectation $\mathscr{E}[g(X)]$ exists and is given by

$$\mathscr{E}[g(X)] = \sum_j p_j g(x_j) \qquad (4.2.2)$$

if, and only if, this series is absolutely convergent.

Under the assumptions of Theorem 4.2.1 $Y = g(X)$ is a discrete random variable, the values $g(x_j)$ are the possible values of Y. It can happen that several values x_j yield the same value of Y. Let $\{y_k\}$ be the countable set of distinct values of Y, and let $A_k = \{x_j : g(x_j) = y_k\}$. The probabilities

$$q_k = P(Y = y_k) = \sum_{x_j \in A_k} p_j \qquad (4.2.3)$$

determine—together with the set $\{y_k\}$—the distribution function of Y. The expected value of Y exists if, and only if, the series $\sum_k y_k q_k$ is absolutely convergent. We have

$$\sum_k y_k q_k = \sum_k y_k \sum_{x_j \in A_k} p_j = \sum_k \sum_{x_j \in A_k} y_k p_j = \sum_k \sum_{x_j \in A_k} g(x_j) p_j = \sum_j g(x_j) p_j.$$

Therefore

$$\sum_k y_k q_k = \sum_j g(x_j) p_j$$

and the theorem follows.

The situation is more complicated if X is a random variable with an absolutely continuous distribution function $F_X(x)$ and frequency function $p_X(x) = F_X'(x)$. We discuss here only a particular case in detail. Assume that the function $g(x)$ is differentiable and monotone. We consider first the case where $g'(x) > 0$ for all x so that $g(x)$ is strictly monotone increasing. Then

$$y = g(x) \tag{4.2.4a}$$

has a single-valued inverse,

$$x = h(y). \tag{4.2.4b}$$

Since $\mathfrak{M}$ includes all continuous functions $g \in \mathfrak{M}$, $Y = g(X)$ is a random variable. Our assumption does, however, not yet guarantee the existence of $\mathscr{E}(Y)$. To simplify matters we make an assumption that is too restrictive, namely we suppose that

$$\lim_{y \to -\infty} g(y) = a, \quad \lim_{y \to \infty} g(y) = b \quad (-\infty < a < b < +\infty).$$

Then $g(y)$ is bounded and $\mathscr{E}(Y)$ exists by Theorem 4.1.1.

We determine next the distribution function of Y, $F_Y(y)$. We first note that $F_Y(y) = 0$ if $y < a$ and $F_Y(y) = 1$ if $y > b$. Let $a < y < b$, by definition

$$F_Y(y) = P(g(X) \le y) = P(X \le h(y)) = F_X(h(y))$$

and we see that $F_Y(y)$ is also an absolutely continuous distribution function. The frequency function of Y is therefore

$$p_Y(y) = \begin{cases} p_X(h(y)) \dfrac{dh}{dy} & \text{if } a < y < b, \\ 0 & \text{otherwise.} \end{cases} \tag{4.2.5}$$

4.2 EXPECTATIONS OF FUNCTIONS

It follows then that

$$\mathscr{E}(Y) = \int_a^b y p_Y(y)\, dy = \int_a^b y p_X(h(y)) \frac{dh}{dy}\, dy$$

and we see from (4.2.4a) and (4.2.4b) that

$$\mathscr{E}(Y) = \int_{-\infty}^{\infty} g(x) p_X(x)\, dx.$$

We can use the same reasoning if $g'(x) < 0$. In this case, we must replace $\dfrac{dh}{dy}$ by $\left|\dfrac{dh}{dy}\right|$ in (4.2.5). We summarize our result.

Theorem 4.2.2

Let X be an absolutely continuous random variable with frequency function $p_X(x)$ and suppose that $g(x)$ is a strictly monotone function that is differentiable. Assume further that $\lim_{y \to -\infty} g(y) = a$ while $\lim_{y \to \infty} g(y) = b$ (a, b finite). Then $Y = g(X)$ is a random variable and has the expectation

$$\mathscr{E}(Y) = \int_{-\infty}^{\infty} g(x) p_X(x)\, dx.$$

Theorem 4.2.2 is not completely analogous to Theorem 4.2.1, since it imposes some restrictions on $g(x)$ that are not present in Theorem 4.2.1. These restrictions are unnecessary; the following result holds and corresponds completely to Theorem 4.2.1:

Theorem 4.2.3

Let X be a random variable that has an absolutely continuous distribution function $F_X(x)$ and a frequency function $p_X(x)$. Let $g(x) \in \mathfrak{M}$; the expectation of $g(x)$ exists and

$$\mathscr{E}[g(X)] = \int_{-\infty}^{\infty} g(x) p_X(x)\, dx$$

if, and only if, the integral

$$\int_{-\infty}^{\infty} |g(x)| p_X(x)\, dx$$

is convergent.

This result can be extended to the multivariate case. Let $X_1, \ldots, X_n$ be n random variables and let $g(x_1, \ldots, x_n)$ be a function such that $g(X_1, \ldots, X_n)$ is a random variable. This is again a mild restriction on the function g and we denote the set of all functions of n variables having this property by $\mathfrak{M}_n$.

Theorem 4.2.4

Let $X_1, \ldots, X_n$ be n random variables that have an absolutely continuous joint distribution with frequency function $p_{X_1 \cdots X_n}(x_1, \ldots, x_n)$ and let $g(x_1, \ldots, x_n) \in \mathfrak{M}_n$. Then $\mathscr{E}[g(X_1, \ldots, X_n)]$ exists and

$$\mathscr{E}[g(X_1, \ldots, X_n)]$$
$$= \int_{-\infty}^{\infty} \cdots \int_{-\infty}^{\infty} g(x_1, \ldots, x_n) p_{X_1 \cdots X_n}(x_1, \ldots, x_n)\, dx_1 \cdots dx_n$$

if, and only if, the integral

$$\int_{-\infty}^{\infty} \cdots \int_{-\infty}^{\infty} |g(x_1, \ldots, x_n)| p_{X_1 \cdots X_n}(x_1, \ldots, x_n)\, dx_1 \cdots dx_n$$

is convergent.

The proofs of Theorems 4.2.3 and 4.2.4 would exceed the scope of this book† and are therefore omitted.

---

† See [*1*, pp. 87–89] for proof of Theorems 4.2.3 and 4.2.4.

Theorem 4.2.5

Let $X_1, X_2, \ldots, X_n$ be n discrete random variables with a countable set of possible values $\{x_{1,i_1}, x_{2,i_2}, \ldots, x_{n,i_n}\}$ and let $p_{i_1 \cdots i_n} = P(X_1 = x_{1,i_1}, \ldots, X_n = x_{n,i_n})$. Then

$$\mathscr{E}[g(X_1, \ldots, X_n)] = \sum_{i_1} \cdots \sum_{i_n} g(x_{i_1}, \ldots, x_{i_n}) p_{i_1 \cdots i_n} \qquad (4.2.6)$$

if, and only if, the sum in (4.2.6) is absolutely convergent.

The proof of Theorem 4.2.5 is carried along the lines of the proof of Theorem 4.2.1.

4.3 Properties of Expectations

Theorem 4.3.1 (Addition Rule for Expectations)

Let X and Y be two random variables and assume that X and Y are either both discrete or both absolutely continuous. Suppose further that the expectations $\mathscr{E}(X)$ and $\mathscr{E}(Y)$ exist, then $\mathscr{E}(X + Y)$ exists and $\mathscr{E}(X + Y) = \mathscr{E}(X) + \mathscr{E}(Y)$.

Proof

First we prove the theorem for the discrete case. Let $\{x_i\}$ and $\{y_j\}$ be the possible values of X and Y, respectively, and let $p_{ij} = P(X = x_i, Y = y_j)$. The possible values of $X + Y$ are $x_i + y_j$ and we note that

$$\sum_i \sum_j |x_i + y_j| p_{ij} \leq \sum_i \sum_j |x_i| p_{ij} + \sum_i \sum_j |y_j| p_{ij}$$
$$= \sum_i |x_i| p_{i\cdot} + \sum_j |y_j| p_{\cdot j}.$$

We see therefore that the expected value of $X + Y$ exists, and we obtain

$$\sum_i \sum_j (x_i + y_j) p_{ij} = \sum_i x_i p_{i\cdot} + \sum_j y_j p_{\cdot j} = \mathscr{E}(X) + \mathscr{E}(Y)$$

so that the statement is proven for discrete X and Y.

If X and Y are both absolutely continuous, we see that

$$\int_{-\infty}^{\infty} \int_{-\infty}^{\infty} |x + y| \, p_{XY}(x, y) \, dx \, dy \leq \int_{-\infty}^{\infty} |x| p_X(x) \, dx + \int_{-\infty}^{\infty} |y| p_Y(y) \, dy$$

so that the expectation exists and the theorem follows in the same way as in the discrete case.

The result can immediately be extended to finite sums of random variables by induction. Using Theorem 4.1.3, one obtains the following corollary

Corollary to Theorem 4.3.1

Let $X_1, \ldots, X_n$ be n random variables and let $a_1, \ldots, a_n$ be n real constants. Suppose that $\mathscr{E}(X_j)$ exists for $j = 1, \ldots, n$; then

$$\mathscr{E}\left[\sum_{j=1}^n a_j X_j\right] = \sum_{j=1}^n a_j \mathscr{E}(X_j).$$

We consider next the product of independent random variables.

Theorem 4.3.2 (Multiplication Rule for Expectations)

Let X and Y be two random variables and assume that both are either discrete or absolutely continuous. Suppose further that $\mathscr{E}(X)$ and $\mathscr{E}(Y)$ exist and that X and Y are independent. Then the expectation of the product of X and Y exists and $\mathscr{E}(XY) = \mathscr{E}(X)\mathscr{E}(Y)$.

4.3 PROPERTIES OF EXPECTATIONS

Proof

We consider first the case where X and Y are both absolutely continuous. Let $p_{XY}(x, y)$ be their joint frequency function. It follows from the independence of X and Y that

$$p_{XY}(x, y) = p_X(x) p_Y(y).$$

Therefore

$$\int_{-\infty}^{\infty} \int_{-\infty}^{\infty} |xy| p_{XY}(x, y) \, dx \, dy = \left[\int_{-\infty}^{\infty} |x| p_X(x) \, dx \right] \left[\int_{-\infty}^{\infty} |y| p_Y(y) \, dy \right]$$

and we see that the expectation of XY exists, and obtain by a similar argument the statement of the theorem.

The proof is carried in the same way if X and Y are both discrete and is left to the reader.

It is again possible to extend the theorem to any finite number of random variables.

Remark

The expectation of the sum equals the sum of the expectations of the summands without assuming their independence. For the validity of the multiplication rule, the assumption of independence of the factors is essential.

A random variable X is said to be nonnegative if $P(X \geq 0) = 1$ or—equivalently—if $P(X < 0) = 0$.

Theorem 4.3.3

The expectation of a nonnegative random variable is nonnegative.

We give the proof only for the absolutely continuous case.

Proof

Suppose that X is a nonnegative random variable, then $p_X(x) = 0$ for $x < 0$, so that

$$\mathscr{E}(X) = \int_{-\infty}^{\infty} x p_X(x)\, dx = \int_{0}^{\infty} x p_X(x)\, dx \geq 0$$

provided that the expectation of X exists.

4.4 Moments

In Section 4.3 we discussed the expected values of functions of random variables. All powers of x with positive-integer exponents are continuous functions and belong therefore to $\mathfrak{M}$, their mathematical expectations are often useful.

Let X be a random variable and let k be a positive integer and suppose that the expectation of X^k exists. Then

$$\mathscr{E}(X^k) = \alpha_k \tag{4.4.1}$$

is called the moment (algebraic moment) of order k of the random variable X. Similarly,

$$\mathscr{E}(|X|^k) = \beta_k \tag{4.4.2}$$

is called the absolute moment of order k of X. Our definition of the expectation was formulated in such a manner that the assumptions of the existence of the kth moment and of the kth absolute moment are equivalent. The expectation

$$\mu_k = \mathscr{E}[(X - \alpha_1)^k] \tag{4.4.3}$$

is called the central moment of order k of the random variable X.

Theorem 4.4.1

Let X be a random variable and assume that $\mathscr{E}(X^k)$ exists, then $\mathscr{E}(X^r)$ exists for $r < k$.

We prove the statement for the absolutely continuous case.

4.4 MOMENTS

Proof

We have

$$\int_{-\infty}^{\infty} |x|^r p_X(x)\, dx = \int_{-1}^{+1} |x|^r p_X(x)\, dx + \int_{|x|>1} |x|^r p_X(x)\, dx.$$

The first integral is always finite. Since $|x|^r < |x|^k$ for $r < k$ and $|x| \geq 1$, it follows from the assumptions of the theorem that the second integral exists so that the statement is proven. In the discrete case the proof is carried in the same way and is therefore omitted.

The expected value of a random variable is sometimes called a measure of location. This term can be explained by the following physical interpretation of a distribution: In the case of a univariate discrete distribution a unit mass is distributed over all discontinuity points of the distribution in such a way that the mass p_j is placed at the discontinuity point x_j. In the case of an absolutely continuous distribution the unit mass is distributed over the real axis in such a way that each interval, however small, should carry the mass assigned to it by the distribution function. The expected value of the random variable becomes then the center of gravity of the "probability mass" and, in a way, gives an idea of its location. However, the expectation does not give any information on the manner in which the probability mass is scattered about its center of gravity.

The second central moment is a typical value that indicates how the probability mass is scattered about its center of gravity.† In our physical model it corresponds to the moment of inertia of the probability mass.

---

† This is, of course, not the only typical value that is suitable for this purpose. The first absolute central moment $\mathscr{E}[\,|X - \mathscr{E}(X)|\,]$ or any central moment of even order could be used instead. The avoidance of absolute values and of higher powers simplifies the computations so that the second central moment is the most useful measure for the scatter (dispersion) of a distribution. Similarly, it is also possible to construct measures of location different from $\mathscr{E}(X)$. We mention here only the median of X. The number m is said to be a median of X if $F_X(m - 0) \leq \tfrac{1}{2} \leq F_X(m)$. A median divides the probability mass into two equal parts. A random variable may have more than one median.

Let X be a random variable, the variance of X, in symbols $\text{Var}(X)$ or σ_X^2, is defined as
$$\text{Var}(X) = \mathscr{E}\{[X - \mathscr{E}(X)]^2\} = \mu_2$$
provided that the expected value on the right-hand side of this formula exists.

It follows immediately from this definition that
$$\text{Var}(X) = \mathscr{E}(X^2) - [\mathscr{E}(X)]^2. \tag{4.4.4}$$
Formula (4.4.4) indicates that the existence of the second moment is a necessary and sufficient condition for the existence of the variance.

Let X be a random variable which has a finite second moment,† the quantity
$$\sigma_X = \sqrt{\text{Var}(X)} \tag{4.4.5}$$
is called the standard deviation of X. The square root is taken with the positive sign in (4.4.5).

Let c be an arbitrary constant, it follows from (4.4.4) that
$$\mathscr{E}[(X - c)^2] - \text{Var}(X) = [\mathscr{E}(X) - c]^2. \tag{4.4.6}$$

Theorem 4.4.2

Let X be a random variable that has a finite second moment and let c be a real constant. Then
$$\mathscr{E}[(X - c)^2] \geq \text{Var}(X).$$
Moreover, the expectation $\mathscr{E}[(X - c)^2]$ attains its minimum if $c = \mathscr{E}(X)$.

Example 4.4.1

Let X be the number of points obtained in a single toss of a die. We found $\mathscr{E}(X) = \frac{7}{2}$ (Example 4.1.2). We have
$$\mathscr{E}(X^2) = \sum_{j=1}^{6} j^2 \frac{1}{6} = \frac{91}{6} \quad \text{and} \quad \text{Var}(X) = \frac{91}{6} - \frac{49}{4} = \frac{35}{12}.$$

---

† The statement "X has a finite second moment" means that the second moment of X exists.

4.4 MOMENTS

Example 4.4.2

Let X be a random variable that has a geometric distribution. We compute the second moment $\alpha_2 = p \sum_{j=1}^{\infty} j^2 q^{j-1}$. We have (see Example 4.1.3)

$$\alpha_1 = \mathcal{E}(X) = \frac{1}{p}, \qquad \frac{d}{dq} \sum_{j=1}^{\infty} q^j = (1 - q)^{-2}$$

so that

$$\sum_{j=1}^{\infty} jq^j = q \frac{d}{dq} \sum_{j=1}^{\infty} q^j = q(1 - q)^{-2}.$$

Therefore

$$\sum_{j=1}^{\infty} j^2 q^{j-1} = \frac{d}{dq} \sum_{j=1}^{\infty} jq^j = \frac{d}{dq} [q(1-q)^{-2}] = \frac{1+q}{(1-q)^3}.$$

We obtain therefore $\alpha_2 = \dfrac{1+q}{p^2}$ and, in view of (4.4.4), $\text{Var}(X) = \dfrac{q}{p^2}$.

Example 4.4.3

Let X be a random variable that has a normal distribution with parameters α and σ. We already know that $\mathcal{E}(X) = \alpha$ so that

$$\sigma_X^2 = \frac{1}{\sigma\sqrt{2\pi}} \int_{-\infty}^{\infty} (x - \alpha)^2 \exp\left[-\frac{1}{2\sigma^2}(x-\alpha)^2\right] dx$$

$$= \sigma^2 \frac{1}{\sqrt{2\pi}} \int_{-\infty}^{\infty} y^2 \exp\left(-\frac{y^2}{2}\right) dy.$$

We must therefore compute the integral

$$\frac{1}{\sqrt{2\pi}} \int_{-\infty}^{\infty} y^2 \exp\left(-\frac{y^2}{2}\right) dy = \frac{1}{\sqrt{2\pi}} \int_{-\infty}^{\infty} y \cdot \left[y \exp\left(-\frac{y^2}{2}\right) dy\right]$$

$$= \frac{1}{\sqrt{2\pi}} \int_{-\infty}^{\infty} y \, d\left[-\exp\left(-\frac{y^2}{2}\right)\right].$$

Integrating by parts we obtain

$$\frac{1}{\sqrt{2\pi}} \int_{-\infty}^{\infty} y^2 \exp\left(-\frac{y^2}{2}\right) dy$$

$$= \left| -\frac{1}{\sqrt{2\pi}} y \exp\left(-\frac{y^2}{2}\right) \right|_{-\infty}^{\infty} + \frac{1}{\sqrt{2\pi}} \int_{-\infty}^{\infty} \exp\left(-\frac{y^2}{2}\right) dy = 1$$

so that $\sigma_X^2 = \sigma^2$.

Theorem 4.4.3

Let X be a random variable with finite second moment and let a and b be two real constants. Then $\mathrm{Var}(aX + b) = a^2\, \mathrm{Var}(X)$.

This follows immediately from formula (4.4.4).

Let X be a nondegenerate random variable with finite second moment. Consider the random variable $X^* = \dfrac{X - \mathscr{E}(X)}{\sigma_X}$. It follows from the Corollary to Theorem 4.3.1 and from the last theorem that $\mathscr{E}(X^*) = 0$ and $\mathrm{Var}(X^*) = 1$. The random variable X^* is called the standardized variable X.

Theorem 4.4.4

An absolutely continuous random variable cannot have zero variance. A discrete random variable has zero variance if, and only if, it is degenerate.

Let X be absolutely continuous and put $\alpha = \mathscr{E}(X)$. The relation $\mathrm{Var}(X) = \int_{-\infty}^{\infty} (x - \alpha)^2 p(x)\, dx = 0$ can be satisfied only if $p(x) = 0$ whenever $x \neq \alpha$. However, this is not possible for a frequency function. Let X be a discrete random variable and suppose that

$$\mathrm{Var}(X) = \sum_j [x_j - \mathscr{E}(X)]^2 p_j = 0.$$

4.4 MOMENTS

This relation can only be satisfied if $x_j = \mathscr{E}(X)$ for all j for which $p_j > 0$, but this means that X has only the possible value $\mathscr{E}(X)$, and is therefore degenerate. The converse statement, namely that the variance of a degenerate random variable is zero, follows from (4.4.4).

Theorem 4.4.5

Let X and Y be two random variables and suppose that $\mathscr{E}(X^2)$ and $\mathscr{E}(Y^2)$ exist. Then $\mathscr{E}(XY)$ exists and

$$[\mathscr{E}(XY)]^2 \leq \mathscr{E}(X^2)\mathscr{E}(Y^2). \tag{4.4.7}$$

The first part of the statement follows from the elementary inequality†

$$|ab| \leq \tfrac{1}{2}(a^2 + b^2).$$

To prove the second part of the statement we note that the random variable $(tX + Y)^2$ is nonnegative for any real value of t, therefore (Theorem 4.3.3)

$$\mathscr{E}[(tX + Y)^2] = t^2 \mathscr{E}(X^2) + 2t\mathscr{E}(XY) + \mathscr{E}(Y^2) \geq 0.$$

Then

$$t^2 \mathscr{E}(X^2) + 2t\mathscr{E}(XY) + \mathscr{E}(Y^2) = 0$$

is a quadratic equation in t, which has either no real roots or a double root; therefore $[\mathscr{E}(XY)]^2 - \mathscr{E}(X^2)\mathscr{E}(Y^2) \leq 0$ and (4.4.7) follows. Relation (4.4.7) is called Schwarz's inequality.

Let X and Y be two independent random variables and assume that $\mathscr{E}(X^2)$ and $\mathscr{E}(Y^2)$ exist. According to (4.4.4) we have

$$\begin{aligned}\operatorname{Var}(X + Y) &= \mathscr{E}(X^2) + 2\mathscr{E}(XY) + \mathscr{E}(Y^2) \\ &\quad - [\mathscr{E}(X)]^2 - 2\mathscr{E}(X)\mathscr{E}(Y) - [\mathscr{E}(Y)]^2.\end{aligned} \tag{4.4.8}$$

---

† This follows from $(|a| - |b|)^2 \geq 0$.

Since X and Y are independent, we see that $\mathscr{E}(XY) = \mathscr{E}(X)\mathscr{E}(Y)$ and conclude from (4.4.8) that for independent X and Y

$$\operatorname{Var}(X + Y) = \operatorname{Var}(X) + \operatorname{Var}(Y). \tag{4.4.9}$$

We extend this to the sum of n independent random variables by induction.

Theorem 4.4.6

Let $X_1, \ldots, X_n$ be n independent random variables and assume that each of these has a finite second moment. Then

$$\operatorname{Var}\left(\sum_{j=1}^{n} X_j\right) = \sum_{j=1}^{n} \operatorname{Var}(X_j).$$

We combine this result with Theorem 4.4.3 and obtain the

Corollary to Theorem 4.4.6

Let $X_1, \ldots, X_n$ be n independent random variables and let $c_1, \ldots, c_n$ be n constants and suppose that the second moment of each X_j exists. Then

$$\operatorname{Var}\left(\sum_{j=1}^{n} c_j X_j\right) = \sum_{j=1}^{n} c_j^2 \operatorname{Var}(X_j).$$

Let X and Y be two—possibly dependent—random variables and assume that they have finite second moments. We write

$$\alpha_{10} = \mathscr{E}(X), \quad \alpha_{01} = \mathscr{E}(Y) \tag{4.4.10}$$

for their first moments and denote their second moments by

$$\alpha_{20} = \mathscr{E}(X^2), \quad \alpha_{02} = \mathscr{E}(Y^2). \tag{4.4.11a}$$

In the bivariate case one has also a mixed moment (product–moment) of second order

$$\alpha_{11} = \mathscr{E}(XY). \tag{4.4.11b}$$

4.4 MOMENTS

Similarly, one can define central moments of second order

$$\mu_{20} = \mathscr{E}[(X - \alpha_{10})^2], \qquad \mu_{02} = \mathscr{E}[(Y - \alpha_{01})^2],$$
$$\mu_{11} = \mathscr{E}[(X - \alpha_{10})(Y - \alpha_{01})]. \qquad (4.4.12)$$

If one wishes to put the random variables into evidence, one writes σ_X^2, σ_Y^2, and σ_{XY} instead of μ_{20}, μ_{02}, and μ_{11}. The central moments μ_{20} and μ_{02} were introduced earlier; they are the variances of X and Y

$$\mu_{20} = \text{Var}(X) = \sigma_X^2, \qquad \mu_{02} = \text{Var}(Y) = \sigma_Y^2.$$

The mixed central moment μ_{11} of X and Y is called the covariance of X and Y and one often writes

$$\mu_{11} = \sigma_{XY} = \text{Cov}(X, Y).$$

A simple computation shows that

$$\text{Cov}(X, Y) = \alpha_{11} - \alpha_{10}\alpha_{01}. \qquad (4.4.13)$$

Theorem 4.4.7

Let X and Y be two independent random variables with finite second moments. Then $\text{Cov}(X, Y) = 0$.

The converse is, however, not true. Random variables with zero covariance need not be independent.
This is illustrated by the following example.

Example 4.4.4

Let X and Y be two absolutely continuous random variables with frequency function

$$p_{XY}(x, y) = \begin{cases} \frac{1}{4}(1 - x^3 y + xy^3) & \text{if } |x| < 1 \text{ and } |y| < 1, \\ 0 & \text{otherwise.} \end{cases}$$

It is easily seen that $\text{Cov}(X, Y) = 0$. The marginal distributions are given by

$$p_X(x) = \begin{cases} \frac{1}{2} & \text{if } |x| < 1, \\ 0 & \text{otherwise,} \end{cases} \qquad p_Y(y) = \begin{cases} \frac{1}{2} & \text{if } |y| < 1, \\ 0 & \text{otherwise.} \end{cases}$$

Hence $p_X(x)p_Y(y) \neq p_{XY}(x, y)$ so that X and Y are not independent.

Let X and Y be two nondegenerate random variables with finite second moments. The number

$$\rho(X, Y) = \frac{\sigma_{XY}}{\sigma_X \sigma_Y} \tag{4.4.14}$$

is called the coefficient of correlation of X and Y.

The coefficient of correlation of two random variables is zero if, and only if, their covariance is zero. In this case, we say that the random variables are uncorrelated. Independent random variables are always uncorrelated, but the converse (see Example 4.4.4) is not true.

Next we derive bounds for the correlation coefficient.

Let u and v be two arbitrary real variables. Then

$$\mathscr{E}\{[u(X - \alpha_{10}) + v(Y - \alpha_{01})]^2\} = u^2 \sigma_X^2 + 2uv\sigma_{XY} + v^2 \sigma_Y^2 \geq 0. \tag{4.4.15}$$

We use the reasoning employed in the derivation of (4.4.7) and see that

$$\sigma_{XY}^2 \leq \sigma_X^2 \sigma_Y^2$$

so that

$$|\rho(X, Y)| \leq 1. \tag{4.4.16}$$

In order to see what $|\rho| = 1$ means we put $u = \rho \sigma_Y$, $v = -\sigma_X$ in (4.4.15) and obtain

$$\mathscr{E}\{[\rho\sigma_Y(X - \alpha_{10}) - \sigma_X(Y - \alpha_{01})]^2\} = \rho^2 \sigma_X^2 \sigma_Y^2 - 2\rho\sigma_X \sigma_Y \sigma_{XY} + \sigma_X^2 \sigma_Y^2.$$

If $|\rho| = 1$ this becomes

$$\mathscr{E}\{[\rho\sigma_Y(X - \alpha_{10}) - \sigma_X(Y - \alpha_{01})]^2\} = 0.$$

4.4 MOMENTS

Also,

$$\mathcal{E}\{\rho\sigma_Y(X - \alpha_{10}) - \sigma_X(Y - \alpha_{01})\} = 0$$

so that the random variable $\rho\sigma_Y(X - \alpha_{10}) - \sigma_X(Y - \alpha_{01})$ has the degenerate distribution function $\varepsilon(x)$. This means that the random variables X and Y are linearly dependent and satisfy the relation

$$\rho\sigma_Y(X - \alpha_{10}) - \sigma_X(Y - \alpha_{01}) = 0. \qquad (4.4.17)$$

Let us conversely assume that there exists a linear relation

$$aX + bY = c \qquad (4.4.18)$$

between the nondegenerate random variables X and Y, where $a \neq 0$, $b \neq 0$. Taking expectations and subtracting the relation so obtained from (4.4.18) we see that

$$a(X - \alpha_{10}) + b(Y - \alpha_{01}) = 0.$$

We put $u = a$ and $v = b$ in (4.4.15) and see that

$$a^2\sigma_X^2 + 2ab\sigma_{XY} + b^2\sigma_Y^2 = 0$$

or, using the relation $\sigma_{XY} = \rho\sigma_X\sigma_Y$,

$$(1 - \rho^2)a^2\sigma_X^2 + (\rho\sigma_X a + b\sigma_Y)^2 = 0.$$

This is only possible if $\rho^2 = 1$, we see therefore that $|\rho(X, Y)| = 1$ means that X and Y are linearly dependent.

Theorem 4.4.8

Let $\rho(X, Y)$ be the coefficient of correlation between two nondegenerate random variables X and Y. Then $|\rho(X, Y)| = 1$ if, and only if, there exists a linear relation with nonzero coefficients between X and Y.

In Section 3.5, we computed the first moments α_{10} and α_{01} as well as the variances σ_1^2 and σ_2^2 of the bivariate normal distribution. We shall now compute the covariance of X and Y.

We have

$$\text{Cov}(X, Y) = \mathscr{E}[(X - \alpha_{10})(Y - \alpha_{01})]$$

$$= \int_{-\infty}^{\infty} \int_{-\infty}^{\infty} (x - \alpha_{10})(y - \alpha_{01}) p(x, y) \, dx \, dy,$$

or using (3.5.17),

$$\text{Cov}(X, Y) = \frac{1}{2\pi\sigma_1\sigma_2\sqrt{1 - \rho^2}} \int_{-\infty}^{\infty} \int_{-\infty}^{\infty} uv \exp\left[-\frac{1}{2} Q(u, v)\right] du \, dv,$$

where

$$Q(u, v) = \frac{1}{1 - \rho^2}\left(\frac{u^2}{\sigma_1^2} - \frac{2\rho uv}{\sigma_1\sigma_2} + \frac{v^2}{\sigma_2^2}\right).$$

An elementary computation shows that

$$Q(u, v) = \frac{u^2}{\sigma_1^2} + \frac{1}{1 - \rho^2}\left(\frac{v}{\sigma_2} - \frac{\rho u}{\sigma_1}\right)^2. \tag{4.4.19}$$

First we compute the inner integral

$$\mathscr{T}(u) = \int_{-\infty}^{\infty} v \exp[-\tfrac{1}{2} Q(u, v)] \, dv$$

and we see from (4.4.19) that

$$\mathscr{T}(u) = \exp\left(-\frac{u^2}{2\sigma_1^2}\right) \int_{-\infty}^{\infty} v \exp\left[-\frac{1}{2(1 - \rho^2)}\left(\frac{v}{\sigma_2} - \frac{\rho u}{\sigma_1}\right)^2\right] dv. \tag{4.4.20}$$

Substituting $z = \dfrac{1}{\sqrt{1 - \rho^2}}\left(\dfrac{v}{\sigma_2} - \dfrac{\rho u}{\sigma_1}\right)$ into (4.4.20) we see that

$$\mathscr{T}(u) = \exp\left(-\frac{u^2}{2\sigma_1^2}\right) u \frac{\rho\sigma_2^2\sqrt{2\pi(1 - \rho^2)}}{\sigma_1}.$$

Since

$$\text{Cov}(X, Y) = \frac{1}{2\pi\sigma_1\sigma_2\sqrt{1 - \rho^2}} \int_{-\infty}^{\infty} u\mathscr{T}(u) \, du,$$

4.4 MOMENTS

an elementary computation, yields

$$\text{Cov}(X, Y) = \rho\sigma_1\sigma_2. \qquad (4.4.21)$$

This gives us an interpretation of the parameter ρ introduced in (3.5.15b) namely

$$\rho = \frac{\sigma_{XY}}{\sigma_X \sigma_Y}.$$

The parameter ρ is called the coefficient of correlation of the random variables X and Y.

Theorem 4.4.9 (Chebyshev's Inequality)

Let X be a random variable and suppose that the variance σ^2 of X exists. Then

$$P(|X - \mathscr{E}(X)| \geq k\sigma) \leq \frac{1}{k^2},$$

where k is an arbitrary positive number.

We prove the theorem for the absolutely continuous case.

$$\sigma^2 = \int_{-\infty}^{\infty} [x - \mathscr{E}(X)]^2 p(x)\, dx \geq \int_{|x - \mathscr{E}(X)| \geq k\sigma} [x - \mathscr{E}(X)]^2 p(x)\, dx$$
$$\geq k^2 \sigma^2 P(|X - \mathscr{E}(X)| \geq k\sigma)$$

and the statement follows immediately. The proof for the discrete case is carried in the same way.

Remark 1

Chebyshev's inequality is valid without any assumption concerning the form of the distribution of X [only the existence of Var(X) is needed] and is therefore applicable to a wide variety of distributions.

Remark 2

It can be shown that Chebyshev's inequality is not only valid for absolutely continuous and discrete distributions, but also for random variables that do not fall into these two special classes, provided that their variance exists.

It is often convenient to write Chebyshev's inequality in a slightly different form by putting $k = \dfrac{\varepsilon}{\sigma}$; one obtains then the inequality

$$P(|X - \mathscr{E}(X)| \geq \varepsilon) \leq \frac{\operatorname{Var}(X)}{\varepsilon^2}. \qquad (4.4.22)$$

Example 4.4.5

We consider the experiment of tossing a coin for which the probability of head in a single trial is unknown. Let p be this probability. We wish to solve the following problem: How many trials must one make in order to assure with a probability of 95% that the relative frequency of heads does not differ from p by more than $\tfrac{1}{10}$?

Let S_n be the observed number of heads in n tosses; then $\dfrac{S_n}{n}$ is the relative frequency of heads in n trials. We wish to determine n so that

$$P\left(\left|\frac{S_n}{n} - p\right| < \frac{1}{10}\right) \geq \frac{95}{100} \text{ or, equivalently } P\left(|S_n - np| \geq \frac{n}{10}\right) \leq \frac{5}{100}.$$

Since $\operatorname{Var}(S_n) = np(1-p) \leq \dfrac{n}{4}$ we see from (4.4.22) that

$$P\left(|S_n - np| \geq \frac{n}{10}\right) \leq \frac{np(1-p)}{n^2/100} \leq \frac{25}{n}.$$

We must therefore select n so that $\dfrac{25}{n} \leq \dfrac{5}{100}$ or $n \geq 500$.

4.5 Regression

Theorem 4.4.8 indicates that the coefficient of correlation describes some features of the dependence of two random variables. A more detailed idea of the dependence of two random variables is obtained by studying the conditional frequency functions, or, in the discrete case, the conditional probabilities. We discuss here only the absolutely continuous case; the discrete case can be treated in an analogous manner. Let X and Y be two absolutely continuous random variables and let

4.5 REGRESSION

$p_X(x|y) = p_X(x|Y = y)$ be the conditional frequency function of X, given that $Y = y$ and $p_Y(y|x) = p_Y(y|X = x)$ be the conditional frequency function of Y, given that $X = x$. Since these are frequency functions, one can define the conditional expectation

$$\mathscr{E}(Y|X) = \mathscr{E}(Y|x) = \int_{-\infty}^{\infty} y p_Y(y|x) \, dy \qquad (4.5.1)$$

provided that the integral is absolutely convergent. $\mathscr{E}(Y|X)$ is a random variable, since it depends on the value x that X assumes. We can regard it as a function of x, this function is called the regression curve of Y on X. Similarly, one can define

$$\mathscr{E}(X|Y) = \mathscr{E}(X|y) = \int_{-\infty}^{\infty} x p_X(x|y) \, dx, \qquad (4.5.2)$$

this function is called the regression of X on Y.

As an example, we consider the bivariate normal distribution. We see from (3.5.19a) and (3.5.19b) that

$$\mathscr{E}(Y|x) = \alpha_{01} + \frac{\rho \sigma_Y}{\sigma_X}(x - \alpha_{10}),$$

$$\mathscr{E}(X|y) = \alpha_{10} + \frac{\rho \sigma_X}{\sigma_Y}(y - \alpha_{01}).$$

The regression curves of the bivariate normal distribution are straight lines. The coefficients

$$\begin{aligned} \beta_{2,0} &= \alpha_{01} - \alpha_{10}\frac{\rho \sigma_Y}{\sigma_X}, \\ \beta_{2,1} &= \frac{\rho \sigma_Y}{\sigma_X}, \end{aligned} \qquad (4.5.3)$$

and

$$\begin{aligned} \beta_{1,0} &= \alpha_{10} - \alpha_{01}\frac{\rho \sigma_X}{\sigma_Y}, \\ \beta_{1,1} &= \frac{\rho \sigma_X}{\sigma_Y} \end{aligned} \qquad (4.5.4)$$

are called the regression coefficients. Using the notations of (4.5.3) and (4.5.4) we can write the regression lines in the form

$$\mathscr{E}(Y|X) = \beta_{2,0} + \beta_{2,1} X \tag{4.5.5}$$

$$\mathscr{E}(X|Y) = \beta_{1,0} + \beta_{1,1} Y. \tag{4.5.6}$$

The validity of formulas (4.5.3) and (4.5.4) is not restricted to the case of the bivariate normal distribution. We show next that (4.5.5) holds for any bivariate distribution for which the regression of Y on X is linear.

Theorem 4.5.1

Let X and Y be two random variables with finite second moments. Suppose further that Y has linear regression on X, that is,

$$\mathscr{E}(Y|X) = \beta_0 + \beta_1(X - \alpha_{10}).$$

Then

$$\beta_0 = \alpha_{01} = \mathscr{E}(Y), \qquad \beta_1 = \frac{\rho \sigma_Y}{\sigma_X} \tag{4.5.7}$$

where $\alpha_{01} = \mathscr{E}(Y)$, $\alpha_{10} = \mathscr{E}(X)$, $\sigma_X^2 = \mathrm{Var}(X)$, $\sigma_Y^2 = \mathrm{Var}(Y)$, and ρ is the correlation coefficient of X and Y.

According to our assumptions we have

$$\mathscr{E}(Y|X) = \beta_0 + \beta_1(X - \alpha_{10}),$$

and

$$\mathscr{E}(XY|X) = \beta_0 X + \beta_1(X^2 - \alpha_{10} X).$$

Taking expectations with respect to X we obtain the equations

$$\begin{aligned} \alpha_{01} &= \beta_0, \\ \alpha_{11} &= \beta_0 \alpha_{10} + \beta_1 \sigma_X^2, \end{aligned} \tag{4.5.8}$$

4.5 REGRESSION

or, since $\alpha_{11} - \alpha_{01}\alpha_{10} = \text{Cov}(X, Y) = \rho\sigma_X\sigma_Y$, $\beta_0 = \alpha_{01}$, $\beta_1 = \dfrac{\rho\sigma_Y}{\sigma_X}$ in accordance with the statement of the theorem.

Exchanging X and Y one obtains a result similar to Theorem 4.5.1, which is valid if X has linear regression on Y.

There is an alternative, completely different approach to the problem of regression.

We can determine a straight line $y = \beta_0 + \beta_1 x$ such that

$$H(\beta_0, \beta_1) = \mathscr{E}\{[Y - \beta_0 - \beta_1 X]^2\}$$

is minimized. This procedure of finding the coefficients β_0 and β_1 is called fitting the line by the method of least squares. Since

$$H(\beta_0, \beta_1) = \mathscr{E}\{[(Y - \alpha_{01}) - (\beta_0 - \alpha_{01} + \beta_1\alpha_{10}) - \beta_1(X - \alpha_{10})]^2\},$$

we see that

$$H(\beta_0, \beta_1) = \sigma_Y^2 + (\beta_0 - \alpha_{01} + \beta_1\alpha_{10})^2 + \beta_1^2\sigma_X^2 - 2\beta_1\sigma_{XY}.$$

Therefore

$$\begin{aligned}\dfrac{\partial H}{\partial \beta_0} &= 2(\beta_0 - \alpha_{01} + \beta_1\alpha_{10}), \\ \dfrac{\partial H}{\partial \beta_1} &= 2\alpha_{10}(\beta_0 - \alpha_{01} + \beta_1\alpha_{10}) + 2\beta_1\sigma_X^2 - 2\sigma_{XY}.\end{aligned} \quad (4.5.9)$$

To find the extremum of $H(\beta_0, \beta_1)$ we set the right-hand side of (4.5.9) equal to zero and obtain the equations

$$\beta_0 - \alpha_{01} + \beta_1\alpha_{10} = 0,$$

and

$$\beta_1\sigma_X^2 - \sigma_{XY} = 0.$$

These equations have the solution

$$\beta_0 = \alpha_{01} - \dfrac{\rho\sigma_Y}{\sigma_X}\alpha_{10}; \qquad \beta_1 = \dfrac{\rho\sigma_Y}{\sigma_X}. \quad (4.5.10)$$

It is easy to verify that these values of the coefficients minimize the function $H(\beta_0, \beta_1)$.

We note that the solution (4.5.10) agrees with formula (4.5.7). The straight line determined by the method of least squares is therefore identical with the regression curve, if the regression of Y on X is linear.

The reasoning just given can be repeated with the variables X and Y interchanged.

The theory of regression can be generalized to the case where the regression curve is not a straight line but a parabola of degree s.

We say that the random variable Y has polynomial regression of degree s on X if

$$\mathscr{E}(Y|X) = \beta_0 + \beta_1 X + \cdots + \beta_s X^s. \qquad (4.5.11)$$

The regression coefficients $\beta_0, \beta_1, \ldots, \beta_s$ can be determined by multiplying (4.5.11) by X^j ($j = 0, 1, \ldots, s$). Taking expectations with respect to X one obtains a system of $s + 1$ linear equations for the regression coefficients. It can again be shown that the application of the method of least squares leads to the same regression coefficients.

4.6 Problems

1. Suppose that two unbiased dice are thrown simultaneously. Find the expected value of the sum of points on both dice.
2. A discrete random variable X has all positive integers as its possible values and $P(X = j) = \dfrac{1}{\ln 2} \dfrac{1}{j 2^j}$. Find $\mathscr{E}(X)$.
3. A random variable has a binomial distribution. Find its mean.
4. A random variable has a Poisson distribution. Find its mean.
5. A random variable has a rectangular distribution. Find its mean.
6. A random variable has a Laplace distribution. Find its mean.
7. A random variable has an exponential distribution. Find its mean.

4.6 PROBLEMS

8. Let X be a random variable which assumes the values 2, 3, ... with probabilities $P(X = j) = (j - 1)p^2 q^{j-2}$ ($j = 2, 3, \ldots$, ad infinitum). Find the expectation of X.

9. Let X be a random variable and let k be a positive integer. Suppose that X assumes the values $k + j$ ($j = 0, 1, 2, \ldots$, ad infinitum), with probabilities $P(X = k + j) = \binom{j + k - 1}{k - 1} p^k q^j$ ($j = 0, 1, 2, \ldots$, ad infinitum). Find the expectation of X.

10. Let X be a random variable that has a hypergeometric distribution. Find its expectation.

11. Prove Theorem 4.2.5 for the bivariate case.

12. Prove Theorem 4.3.2 for discrete random variables.

13. Prove Theorem 4.4.1 for the discrete case.

14. Determine the variance of the indicator variable of a set A.

15. Determine the absolute moment of order 1 for a normal distribution with zero mean and standard deviation σ.

16. Suppose that the random variable X has the distribution

$$F(x) = \begin{cases} 0 & \text{for } x < 1, \\ 1 - x^{-m} & \text{for } x \geq 1; \end{cases}$$

find the moments of order k of X. For what values of k do these moments exist?

17. Let

$$p_X(x) = \begin{cases} 0 & \text{if } |x| > r, \\ \dfrac{1}{2r} & \text{if } |x| < r \end{cases}$$

be the frequency function of the random variable X. Determine the moments and absolute moments of X.

18. Let X and Y be two not necessarily independent random variables and suppose that the second moments of X and Y exist. Find the variance of $X + Y$.

19. Let X and Y be two absolutely continuous random variables with joint frequency function
$$p_{XY}(x, y) = \begin{cases} \frac{1}{2} & \text{if } |x+y| \le 1 \text{ and } |x-y| \le 1, \\ 0 & \text{otherwise.} \end{cases}$$
Are X and Y independent? Are they uncorrelated?

20. Prove the following generalization of Chebyshev's inequality: Let $g(x) \in \mathfrak{M}$ be a nonnegative function and let X be a random variable and suppose that the expectation $\mathscr{E}[g(X)]$ exists. Let M be a positive real number and let $A' = \{x: g(x) \ge M\}$. Then
$$P[g(X) \ge M] \le \frac{\mathscr{E}[g(X)]}{M}.$$

21. Let X be a nonnegative random variable and suppose that $\mathscr{E}(X)$ exists. Let k be a positive real constant, then $P(X \ge k\alpha_1) \le \frac{1}{k}$.

22. Let the joint frequency function of X and Y be
$$p_{XY}(x, y) = \begin{cases} 2 & \text{if } 0 < x < y \text{ and } 0 < y < 1, \\ 0 & \text{otherwise.} \end{cases}$$
Determine the regression of Y on X.

23. What is the probability that a normally distributed random variable deviates from its mean by more than twice its standard deviation? Compare your answer with the result which you could obtain from Chebyshev's inequality without using the information that the random variable is normally distributed.

24. In a manufacturing process, the probability of producing a nondefective item is p, whereas the probability of producing a defective item is $q = 1 - p$. The cost of producing an item is c and the price of a good item is s. Defective items cannot be sold.
    (a) What is the expected profit of the manufacturer if N items are produced?
    (b) Suppose that $p = \frac{9}{10}$, $s = \$5.00$, $c = \$2.50$, how large a lot must the manufacturer produce in order to assure that he can expect with probability at least equal to $\frac{1}{10}$ that his profit per unit manufactured exceeds $\$1.80$?

(c) Suppose that the manufacturer can choose between the process described in (b), that is with $p_1 = \frac{9}{10}$, $s_1 = \$5.00$, $c_1 = \$2.50$ and a process with $p_2 = \frac{85}{100}$, $s_2 = \$5.00$, $c_2 = \$3.00$. Which of the two processes is more profitable for the manufacturer?

25. Let X be a normally distributed random variable with mean zero and unit variance. Use Chebyshev's inequality to estimate the probability that the absolute value of the difference $X - \mathscr{E}(X)$ will exceed 1.96 times the standard deviation of X. Compare this with the exact value obtained from tables of the normal distribution.

26. Let X be a normally distributed random variable with mean $\alpha > 0$ and variance $\sigma^2 = \left(\frac{\alpha}{3}\right)^2$. Find the probability that X assumes a negative value.

27. Let X be an arbitrary, absolutely continuous random variable and suppose that $\mathscr{E}(\exp X^2)$ exists. Show that $P(|X| \geq \varepsilon) \leq \dfrac{\mathscr{E}(\exp X^2)}{\exp \varepsilon^2}$.

28. Suppose that the random variable X has the frequency function

$$p(x) = \begin{cases} \dfrac{x^m e^{-x}}{m!} & \text{for } x \geq 0, \\ 0 & \text{for } x < 0. \end{cases}$$

Show that $P(0 < X < 2(m+1)) \geq \dfrac{m}{m+1}$.

29. Suppose that the random variable X has a Poisson distribution and that $\mathscr{E}(X) = 6$. Show that $P(0 < X < 9) > \frac{1}{3}$.

Reference

1. H. Tucker, "An Introduction to Probability and Mathematical Statistics." Academic Press, New York, 1962.

5 Limit Theorems

5.1 Laws of Large Numbers

We mentioned in Section 1.2 that the assignment of probabilities to events was motivated by the wish to reflect the phenomenon of statistical regularity in the mathematical model. In this section, we discuss theorems which indicate that relative frequencies approach in a certain sense the corresponding probabilities, so that our model is justified.

Theorem 5.1.1 (Bernoulli's Law of Large Numbers)

Let $\{X_n\}$ be a sequence of Bernoulli trials with probability of success equal to p and let S_n be the number of successes in n trials. Then for any $\varepsilon > 0$ the relation $\lim_{n \to \infty} P\left(\left|\frac{S_n}{n} - p\right| < \varepsilon\right) = 1$ holds.

Proof

First we determine the mean and the variance of S_n. Since X_j is an indicator variable† we have $\mathscr{E}(X_j) = \mathscr{E}(X_j^2) = p$ and $\mathrm{Var}(X_j) = p(1-p)$. We see from Theorems 4.3.1 and 4.4.6 that $\mathscr{E}(S_n) = np$, whereas $\mathrm{Var}(S_n) = np(1-p)$. Therefore $\mathscr{E}\left(\dfrac{S_n}{n}\right) = p$ and $\mathrm{Var}\left(\dfrac{S_n}{n}\right) = \dfrac{p(1-p)}{n}$. We conclude from Chebyshev's theorem that

$$P\left(\left|\frac{S_n}{n} - p\right| \geq \varepsilon\right) \leq \frac{p(1-p)}{n\varepsilon^2} \leq \frac{1}{4\varepsilon^2 n}. \tag{5.1.1}$$

[The last inequality on the right-hand side is a consequence of the fact that $p(1-p) \leq \tfrac{1}{4}$.] We see then that

$$1 \geq P\left(\left|\frac{S_n}{n} - p\right| < \varepsilon\right) \geq 1 - \frac{1}{4\varepsilon^2 n}.$$

The last relation implies the statement of the theorem.

The law of large numbers asserts that Bernoulli trials have the following property: The probability of the relative frequency being close to p approaches 1 as the number of trials increases.

It is also possible to derive a more general result.

Theorem 5.1.2 (Law of Large Numbers)

Let $\{X_n\}$ be a sequence of independent random variables and suppose that the variances $\sigma_n^2 = \mathrm{Var}(X_n)$ exist. Let $S_n = \sum\limits_{j=1}^{n} X_j$ and suppose that

$$\lim_{n \to \infty} \frac{1}{n^2} \sum_{j=1}^{n} \sigma_j^2 = 0.$$

Then

$$\lim_{n \to \infty} P\left(\left|\frac{S_n}{n} - \mathscr{E}\left(\frac{S_n}{n}\right)\right| < \varepsilon\right) = 1$$

for any $\varepsilon > 0$.

---

† Of the set of all ω resulting in success in the jth trial.

5.1 LAWS OF LARGE NUMBERS

Proof

The proof is again carried by Chebyshev's inequality. We have

$$P\left(\left|\frac{S_n}{n} - \mathscr{E}\left(\frac{S_n}{n}\right)\right| \geq \varepsilon\right) \leq \frac{\mathrm{Var}\left(\frac{S_n}{n}\right)}{\varepsilon^2} = \frac{\sum_{j=1}^{n} \sigma_j^2}{\varepsilon^2 n^2}. \tag{5.1.2}$$

According to the assumptions of Theorem 5.1.2 the right-hand side of (5.1.2) tends to zero as n tends to infinity so that the theorem is proven.

Theorem 5.1.2 contains an interesting special case. We say that a sequence of alternatives are a sequence of Poisson trials if the trials are independent and if the probability of success p_j varies from trial to trial. In this case $\sigma_j^2 = p_j(1 - p_j) \leq \frac{1}{4}$ so that $\frac{1}{n^2} \sum_{j=1}^{n} \sigma_j^2 \leq \frac{1}{4n}$. Theorem 5.1.2 can therefore be applied. The probability that the relative frequency of success in Poisson trials does not differ much from its expectation is therefore close to 1, provided that the number of trials is large.

Remark 1

The conditions of Theorem 5.1.2 are satisfied if the X_j are identically distributed random variables with finite variance.

Remark 2

The validity of the law of large numbers is not restricted to sequences of absolutely continuous or discrete random variables.

Remark 3

We say that a sequence X_j of random variables satisfies the law of large numbers if

$$\lim_{n \to \infty} P\left(\frac{|S_n - \mathscr{E}(S_n)|}{n} < \varepsilon\right) = 1$$

for any $\varepsilon > 0$.

5.2 The Central Limit Theorem

The law of large numbers can be used to estimate the probability that S_n lies between two bounds; however, it is often desirable to obtain more precise estimates. The first result in this direction was obtained by De Moivre who investigated the behavior of Bernoulli trials as the number of trials increases indefinitely. In order to discuss this problem we derive first an approximation formula for binomial probabilities. This is of independent interest, since the exact computation of binomial coefficients is very cumbersome.

Let $\{X_j\}$ be a sequence of Bernoulli trials and let p be the probability of success. We write $S_n = \sum_{j=1}^{n} X_j$ and $q = 1 - p$. It is convenient to introduce the random variable

$$S_n^* = \frac{S_n - np}{\sqrt{npq}}. \tag{5.2.1}$$

Clearly $\mathscr{E}(S_n^*) = 0$ while $\mathrm{Var}(S_n^*) = 1$ so that S_n^* is the standardized variable S_n.

Theorem 5.2.1

Let S_n^* be the standardized sum of successes in n Bernoulli trials, then

$$\lim_{n \to \infty} \frac{\sqrt{2\pi npq}\, P(S_n^* = h)}{\exp(-h^2/2)} = 1$$

provided that h remains bounded as n increases.

Proof

The proof of Theorem 5.2.1 uses Taylor's theorem to estimate carefully certain quantities needed in the derivation of the approximation formula for binomial probabilities. The proof is therefore somewhat complicated and may be omitted in a first reading.

5.2 THE CENTRAL LIMIT THEOREM

The binomial probability is given by

$$P(S_n = k) = \binom{n}{k} p^k q^{n-k}. \tag{5.2.2}$$

Stirling's formula can be used to derive an expression for the binomial coefficients (see Appendix A). One obtains

$$\binom{n}{k} = \frac{1}{\sqrt{2\pi n}} \left(\frac{n}{k}\right)^{k+(1/2)} \left(\frac{n}{n-k}\right)^{n-k+(1/2)} e^R, \tag{5.2.3}$$

where

$$R = r(n) - r(k) - r(n-k). \tag{5.2.3a}$$

As n, k, and $n-k$ tend to infinity, R tends to zero. It follows from (5.2.2) and (5.2.3) that

$$P(S_n = k) = \frac{1}{\sqrt{2\pi}} \sqrt{\frac{n}{k(n-k)}} \left(\frac{np}{k}\right)^k \left(\frac{nq}{n-k}\right)^{n-k} e^R. \tag{5.2.4}$$

We put

$$A = \sqrt{2\pi n p q} \, P(S_n = k) \tag{5.2.5}$$

and conclude from (5.2.4) and (5.2.5) that

$$A = \left(\frac{np}{k}\right)^k \left(\frac{nq}{n-k}\right)^{n-k} B e^R, \tag{5.2.6}$$

where

$$B = \sqrt{\frac{n^2 pq}{k(n-k)}}.$$

We let

$$k = np + h\sqrt{npq}, \tag{5.2.7a}$$

so that

$$n - k = nq - h\sqrt{npq}. \tag{5.2.7b}$$

In our computations we will neglect in (5.2.3) the term e^R and obtain an estimate for the binomial coefficient which is valid if n, k and $n-k$ tend to infinity. We see from (5.2.7a) and (5.2.7b) that this condition is satisfied if h remains bounded as n tends to infinity and we note that this is one of the assumptions of Theorem 5.2.1.

We introduce (5.2.7a) and (5.2.7b) into (5.2.6) and take logarithms. In this way we obtain

$$-\log A = (np + h\sqrt{npq}) \log\left(1 + h\sqrt{\frac{q}{np}}\right)$$
$$+ (nq - h\sqrt{npq}) \log\left(1 - h\sqrt{\frac{p}{nq}}\right) - \log B - R. \tag{5.2.8}$$

If n is sufficiently large the expansion

$$\log(1 + x) = x - \tfrac{1}{2}x^2 + \theta x^3 \qquad (|\theta| < 1) \tag{5.2.9}$$

can be applied to the logarithms in (5.2.8) and one obtains

$$-\log A = (np + h\sqrt{npq})\left(h\sqrt{\frac{q}{np}} - \frac{h^2}{2}\frac{q}{np} + \text{terms with } n^{-3/2}\right)$$
$$+ (nq - h\sqrt{npq})\left(-h\sqrt{\frac{p}{nq}} - \frac{h^2}{2}\frac{p}{nq} + \text{terms with } n^{-3/2}\right)$$
$$- \log B - R.$$

A simple computation shows that

$$-\log A = \frac{h^2}{2} + \left(\text{a finite number of terms with common factor } \frac{1}{\sqrt{n}}\right)$$
$$- \log B - R. \tag{5.2.10}$$

We have $B^2 = \dfrac{n^2 pq}{k(n-k)}$ or, substituting (5.2.7a) and (5.2.7b) and simplifying

$$B^2 = \frac{1}{1 + h\dfrac{q-p}{\sqrt{npq}} - \dfrac{h^2}{n}}.$$

5.2 THE CENTRAL LIMIT THEOREM

Hence

$$-2 \log B = \log\left(1 + h \frac{q-p}{\sqrt{npq}} - \frac{h^2}{n}\right)$$

and one sees that $\log B$, and therefore also $-\log B - R$, tends to zero as n tends to infinity. It follows from (5.2.10) that

$$-\log A = \frac{h^2}{2} + \text{an expression that tends to zero as } n \text{ tends to infinity.}$$

(5.2.11)

We see from (5.2.1) that

$$P(S_n = k) = P(np + S_n^* \sqrt{npq} = k)$$

and conclude from (5.2.5) and (5.2.7a) that

$$A = \sqrt{2\pi npq} \, P(S_n^* = h).$$

The statement of Theorem 5.2.1 follows from the last equation and from (5.2.11).

The statement of Theorem 5.2.1 can be reformulated. We put—in accordance with (5.2.7a)—$h = \dfrac{k - np}{\sqrt{npq}}$ and express S_n^* in terms of S_n. One sees that

$$\frac{\sqrt{2\pi npq} \, P(S_n = k)}{\exp\left[-\dfrac{1}{2npq}(k - np)^2\right]}$$

tends to 1 as $n, k,$ and $n - k$ tend to infinity. This means that the binomial probabilities $\binom{n}{k} p^k q^{n-k}$ can be approximated by values of the normal frequency function,

$$\frac{1}{\sigma\sqrt{2\pi}} \exp\left[-\frac{1}{2\sigma^2}(x - \mu)^2\right]$$

by putting $x = k$, $\mu = np$, $\sigma^2 = npq$.

It is often necessary to consider a related problem and to try to approximate the probability that the number S_n of successes in n Bernoulli trials falls within two limits. It is convenient to use the standardized sum S_n^* and to approximate the probability

$$P_n(a, b) = P(a < S_n^* \leq b).$$

This probability is given by

$$P_n(a, b) = \sum_{a < h \leq b} P(S_n^* = h). \tag{5.2.12}$$

Since the possible values of S_n are the integers $0, 1, \ldots, n$ we see that S_n^* can assume any of the $n + 1$ values $h_k = \dfrac{k - np}{\sqrt{npq}}$ $(k = 0, 1, \ldots, n)$ and we note that $h_{k+1} - h_k = \dfrac{1}{\sqrt{npq}}$. Theorem 5.2.1 indicates that the probabilities $P(S_n^* = h)$ can be approximated by $\dfrac{1}{\sqrt{2\pi npq}} \exp\left(-\dfrac{h^2}{2}\right)$. The sum (5.2.12) is therefore similar to a Riemann sum for the integral $\dfrac{1}{\sqrt{2\pi}} \int_a^b \exp\left(-\dfrac{y^2}{2}\right) dy$. This reasoning is only heuristic. A rigorous proof can be carried by means that exceed the scope of this book. We mention therefore only the result without giving a proof.

Theorem 5.2.2 (Central Limit Theorem for Bernoulli Trials)

Let S_n be the number of successes in n Bernoulli trials and put

$$S_n^* = \frac{S_n - np}{\sqrt{npq}}.$$

Then

$$\lim_{n \to \infty} F_{S_n^*}(x) = \frac{1}{\sqrt{2\pi}} \int_{-\infty}^{x} \exp\left(-\frac{y^2}{2}\right) dy.$$

The theorem can be extended to more general situations; we mention here only one result, again without proof.

Theorem 5.2.3 (Central Limit Theorem)†

Let $\{X_j\}$ be a sequence of independently and identically‡ distributed random variables and suppose that their second moments exist. Put $\mathscr{E}(X_j) = \alpha$ and $\mathrm{Var}(X_j) = \sigma^2$ and let

$$S_n^* = \frac{\sum_{j=1}^n X_j - n\alpha}{\sigma\sqrt{n}}$$

then

$$\lim_{n \to \infty} F_{S_n^*}(x) = \frac{1}{\sqrt{2\pi}} \int_{-\infty}^x \exp\left(-\frac{y^2}{2}\right) dy.$$

Remark

It is possible to generalize Theorem 5.2.3 to the case where the X_j are not identically distributed.

5.3 The Poisson Approximation to the Binomial

When we derived Theorem 5.2.1 we assumed that the probability p of success remains constant as n increases so that both np and $n(1-p)$ are large. In this section, we give another approximation to binomial probabilities that is valid if np is small.

We consider a sequence of Bernoulli trials and assume that $np = \lambda$ where λ is a positive constant. Therefore the probability of success depends on the number of trials; it changes from one sequence of trials to the next sequence, since it depends on n, but the probability of success does not vary from trial to trial. We write therefore

$$p_n = \frac{\lambda}{n} \tag{5.3.1}$$

---

† A proof of the central limit theorem is given in Appendix C. It is advisable to read Section 6.1 before reading this appendix. However, studying Appendix C is not necessary for the understanding of the rest of the book. For other proofs of Theorem 5.2.3 see [1, p. 215] and [2, p. 205].

‡ The random variables $\{X_j\}$ are said to be identically distributed if all X_j have the same distribution function.

for the probability of success in the nth sequence of Bernoulli trials. The probability of k successes in the n trials of the nth sequence is

$$P(S_n = k) = \frac{n!}{k!(n-k)!} p_n^k (1 - p_n)^{n-k} \qquad k = (0, 1, \ldots, n),$$

or

$$P(S_n = k) = \frac{n(n-1) \cdots (n-k+1)}{k! n^k} (np_n)^k (1 - p_n)^{n-k}.$$

We substitute for p_n the expression (5.3.1) and obtain

$$P(S_n = k) = \frac{1\left(1 - \frac{1}{n}\right) \cdots \left(1 - \frac{k-1}{n}\right)}{k!} \lambda^k \left[\left(1 - \frac{\lambda}{n}\right)^{n/\lambda}\right]^\lambda \left(1 - \frac{\lambda}{n}\right)^{-k}. \qquad (5.3.2)$$

We see easily that

$$\lim_{n \to \infty} 1\left(1 - \frac{1}{n}\right) \cdots \left(1 - \frac{k-1}{n}\right) = 1,$$

$$\lim_{n \to \infty} \left[\left(1 - \frac{\lambda}{n}\right)^{n/\lambda}\right]^\lambda = e^{-\lambda}, \qquad (5.3.3)$$

$$\lim_{n \to \infty} \left(1 - \frac{\lambda}{n}\right)^{-k} = 1.$$

We write $\pi_k = \lim_{n \to \infty} P(S_n = k)$ and obtain from (5.3.2) and (5.3.3)

$$\pi_k = e^{-\lambda} \frac{\lambda^k}{k!} \qquad (k = 0, 1, \ldots, \text{ad infinitum}). \qquad (5.3.4)$$

The probabilities (5.3.4) are the probabilities of the Poisson distribution that we introduced in Section 3.3 [Example (g)]. We see therefore that the Poisson probabilities can be used as an approximation to the binomial distribution, provided that the conditions stated at the beginning of this section are satisfied.

5.4 Problems

1. Let $a > 2$ and suppose that the random variables X_i ($i = 1, 2, \ldots$) are independently and identically distributed with common distribution function

$$F(x) = \frac{1}{2a^2} \varepsilon(x + a) + \left(1 - \frac{1}{a^2}\right) \varepsilon(x) + \frac{1}{2a^2} \varepsilon(x - a).$$

   Is the law of large numbers valid for the X_i?

2. Let X_k be a random variable that assumes the values $-3^k, 0, 3^k$ with probabilities $P(X_k = -3^k) = P(X_k = 3^k) = 3^{-(2k+2)}$,

$$P(X_k = 0) = 1 - \frac{2}{3^{2k+2}}.$$

   Does the law of large numbers hold for this sequence?

3. We toss a coin and denote by S_n the number of heads in n tosses. How often do we have to toss the coin to assure that with a probability at least equal to 0.99 the frequency of heads deviates from its probability by less than $\frac{1}{100}$? Compare the result obtained by means of Chebyshev's inequality with the result obtained by using the central limit theorem.

4. The deaths of Prussian cavalry soldiers from kicks by horses were recorded during the period from 1875 to 1894. The observed frequencies are given in the following table†:

Number of deaths	Frequency
0	109
1	65
2	22
3	3
4	1

---

† E. Czuber, "Die statistischen Forschungsmethoden." p. 222, L. W. Seidel, Wien, 1927. These data were originally given by L. Bortkiewicz, "Das Gesetz der kleinen Zahlen." B. G. Teubner, Leipzig, 1898.

Compare the observed frequencies with the probabilities obtained by assuming that the number of fatal accidents follows a Poisson distribution with mean value $\lambda = 0.61$.

5. Let X_j ($j = 1, 2, \ldots$) be a sequence of random variables and suppose that X_j has a Poisson distribution with $\mathscr{E}(X_j) = \lambda_j$. Assume further that $\dfrac{1}{n} \sum\limits_{j=1}^{n} \lambda_j = \lambda < \infty$ and show that the law of large numbers holds.

6. Let $\{X_j\}$ be a sequence of independently distributed random variables such that $P(X_j = 2j^s) = P(X_j = -2j^s) = \tfrac{1}{2}$. Find values of s for which the law of large numbers holds.

7. Let $\{X_j\}$ be a sequence of independent random variables such that $P(X_j = j^s) = P(X_j = -j^s) = \tfrac{1}{4}$ while $P(X_j = 0) = \tfrac{1}{2}$. Determine s so that the law of large numbers holds.

8. Let X_k be a random variable that assumes the values $\sqrt{3}, 0$, and $-\sqrt{3}$ with probabilities $\tfrac{1}{6}, \tfrac{2}{3}$, and $\tfrac{1}{6}$, respectively. Suppose that the $\{X_k\}$ are independent. Does this sequence obey the law of large numbers?

9. Let $\{X_k\}$ be a sequence of independently and identically distributed random variables and assume that their common distribution is a uniform distribution. Does the central limit theorem hold for this sequence?

10. Let $\{X_k\}$ be a sequence of independently and identically distributed random variables and assume that their common distribution is an exponential distribution. Does the central limit theorem hold for this sequence?

11. Find an approximate value for the probability that the number of successes in 9000 independent Bernoulli trials with probability $p = \tfrac{1}{3}$ will be contained between 2900 and 3100.

12. Consider a sequence of 50 Bernoulli trials with probability of success $p = 0.1$. Let S_{50} be the number of successes in 50 trials and find the probability $P = P(S_{50} \leq 3)$ of not more than 3 successes:

(a) by exact computation,
   (b) by using the Poisson approximation,
   (c) by using the normal approximation.
13. Consider a sequence of 20 Bernoulli trials with probability of success $p = 0.5$. Let S_{20} be the number of successes in 20 trials and find the probability $P(S_{20} \leq 3)$ by the methods listed in the previous problem.

References

1. H. Cramér, "Mathematical Methods of Statistics." Princeton Univ. Press, Princeton, New Jersey, 1946.
2. H. Tucker, "A Graduate Course in Probability." Academic Press, New York, 1967.

6 | Some Important Distributions

In this chapter we discuss some properties of normally distributed random variables and derive also certain distribution functions that are important in statistics. We need certain tools for this discussion, which we present in the first section of this chapter.

6.1 The Distribution of the Sum of Independent, Absolutely Continuous Random Variables

We first state a theorem that we shall use repeatedly.

Theorem 6.1.1

Let $X_1, \ldots, X_n$ be n random variables whose joint distribution is absolutely continuous. Suppose further that A is a set in the n-dimensional Euclidean space for which $\{\omega : (X_1(\omega), \ldots, X_n(\omega)) \in A\} \in \mathfrak{U}$. Then

$$P[(X_1, \ldots, X_n) \in A] = \int \cdots \int_A p_{X_1 \cdots X_n}(x_1, \ldots, x_n) \, dx_1 \cdots dx_n.$$

Proof

To prove Theorem 6.1.1 we introduce the indicator variable I_A of the set A, that is, we define

$$I_A = I_A(x_1, x_2, \ldots, x_n) = \begin{cases} 1 & \text{if } (x_1, x_2, \ldots, x_n) \in A, \\ 0 & \text{if } (x_1, x_2, \ldots, x_n) \in A^c. \end{cases}$$

Using the argument presented in Example (b) of Section 3.3, we show that $I_A(X_1, X_2, \ldots, X_n)$ is a random variable, which assumes the value 1 with probability $P(A)$ and the value 0 with probability $1 - P(A)$. The expectation $\mathscr{E}(I_A)$ is therefore given by

$$\mathscr{E}(I_A) = P(A).$$

On the other hand one can compute $\mathscr{E}(I_A)$ according to Theorem 4.2.4 and obtain

$$\mathscr{E}(I_A) = \int_{-\infty}^{\infty} \cdots \int_{-\infty}^{\infty} I_A(x_1, x_2, \ldots, x_n) p_{X_1 \cdots X_n}(x_1, \ldots, x_n) \, dx_1 \cdots dx_n.$$

Since $I_A = 0$ outside A, the statement of the theorem follows.

A particular case is often of interest.

Theorem 6.1.2

Let $X_1, \ldots, X_n$ be n absolutely continuous random variables and let $p_{X_1, \ldots, X_n}(x_1, \ldots, x_n)$ be their joint frequency function. Let $g_j(x_1, \ldots, x_n)$ $(j = 1, \ldots, n)$ be n continuous functions and let $A = \{(x_1, \ldots, x_n) : g_1(x_1, \ldots, x_n) \le a_1, \ldots, g_n(x_1, \ldots, x_n) \le a_n\}$. Then $P[(X_1, \ldots, X_n) \in A] = \int \cdots \int_A p_{X_1, \ldots, X_n}(x_1, \ldots, x_n) \, dx_1 \cdots dx_n$.

We use Theorem 6.1.2 to derive the frequency function of the sum of two independent and absolutely continuous random variables.

Theorem 6.1.3

Let X_1 and X_2 be two independent, absolutely continuous random variables and let $p_j(x)$ be the frequency function of X_j ($j = 1, 2$). Let $Z = X_1 + X_2$; the frequency function of Z is then

$$p_Z(x) = \int_{-\infty}^{\infty} p_1(x - v) p_2(v) \, dv.$$

Proof

Since X_1 and X_2 are independent, their joint frequency function is given by $p_1(u) p_2(v)$ and we see from Theorem 6.1.2 that the distribution function of Z is

$$F_Z(x) = P(Z \le x) = P(X_1 + X_2 \le x) = \iint_{u+v \le x} p_1(u) p_2(v) \, du \, dv.$$

It follows that

$$F_Z(x) = \int_{-\infty}^{\infty} \left[p_2(v) \int_{-\infty}^{x-v} p_1(u) \, du \right] dv. \tag{6.1.1}$$

We differentiate (6.1.1) with respect to x and obtain immediately the statement of the theorem.

6.2 Addition of Independent Normal Random Variables

We first consider two independently and normally distributed random variables X_1 and X_2 and admit the possibility that their means and variances are different. We write

$$p_j(x) = \frac{1}{\sigma_j \sqrt{2\pi}} \exp\left[-\frac{1}{2\sigma_j^2} (x - \alpha_j)^2 \right] \quad (j = 1, 2)$$

for the frequency function of X_j. Let $Z = X_1 + X_2$, we conclude from Theorem 6.1.3 that the frequency function $p_Z(x)$ of Z is given by

$$p_Z(x) = \frac{1}{2\pi \sigma_1 \sigma_2} \int_{-\infty}^{\infty} \exp\left[-\frac{1}{2} Q(v) \right] dv,$$

where $Q(v)$ is a polynomial of second degree in v and is given by

$$Q(v) = \frac{(x - v - \alpha_1)^2}{\sigma_1^2} + \frac{(v - \alpha_2)^2}{\sigma_2^2}.$$

A somewhat tedious but quite elementary computation shows that

$$Q(v) = \frac{\sigma_1^2 + \sigma_2^2}{\sigma_1^2 \sigma_2^2} \left[v - \frac{\sigma_1^2 \alpha_2 + \sigma_2^2 (x - \alpha_1)}{\sigma_1^2 + \sigma_2^2} \right]^2 + \frac{(x - \alpha_1 - \alpha_2)^2}{\sigma_1^2 + \sigma_2^2}.$$

Then

$$p_Z(x) = \frac{1}{2\pi \sigma_1 \sigma_2} \exp\left[-\frac{1}{2} \frac{(x - \alpha_1 - \alpha_2)^2}{\sigma_1^2 + \sigma_2^2} \right] \mathcal{T}, \qquad (6.2.1)$$

where

$$\mathcal{T} = \int_{-\infty}^{\infty} \exp\left\{ -\frac{\sigma_1^2 + \sigma_2^2}{2\sigma_1^2 \sigma_2^2} \left[v - \frac{\sigma_1^2 \alpha_2 + \sigma_2^2 (x - \alpha_1)}{\sigma_1^2 + \sigma_2^2} \right]^2 \right\} dv.$$

We introduce a new variable

$$u = \sqrt{\frac{\sigma_1^2 + \sigma_2^2}{\sigma_1^2 \sigma_2^2}} \left[v - \frac{\sigma_1^2 \alpha_2 + \sigma_2^2 (x - \alpha_1)}{\sigma_1^2 + \sigma_2^2} \right]$$

in the integral $\mathcal{T}$ and obtain

$$\mathcal{T} = \frac{\sigma_1 \sigma_2}{\sqrt{\sigma_1^2 + \sigma_2^2}} \int_{-\infty}^{\infty} \exp\left(-\frac{u^2}{2} \right) du,$$

so that (6.2.1) becomes

$$p_Z(x) = \frac{1}{2\pi \sqrt{\sigma_1^2 + \sigma_2^2}} \exp\left[-\frac{1}{2} \frac{(x - \alpha_1 - \alpha_2)^2}{\sigma_1^2 + \sigma_2^2} \right] \cdot \int_{-\infty}^{\infty} \exp\left(-\frac{u^2}{2} \right) du.$$

We know [see formula (4.1.3)] that

$$\frac{1}{\sqrt{2\pi}} \int_{-\infty}^{\infty} \exp\left(-\frac{u^2}{2} \right) du = 1,$$

therefore

$$p_Z(x) = \frac{1}{\sqrt{2\pi(\sigma_1^2 + \sigma_2^2)}} \exp\left[-\frac{1}{2} \frac{(x - \alpha_1 - \alpha_2)^2}{\sigma_1^2 + \sigma_2^2} \right].$$

6.3 THE CHI-SQUARE DISTRIBUTION

This is a normal frequency function with mean $\alpha_1 + \alpha_2$ and variance $\sigma_1^2 + \sigma_2^2$, and we have obtained the following result:

Theorem 6.2.1

Let X_1 and X_2 be two independently and normally distributed random variables and denote the mean and variance of X_j by $\alpha_j = \mathscr{E}(X_j)$ and $\sigma_1^2 = \text{Var}(X_j)$ ($j = 1, 2$). The random variable $X = X_1 + X_2$ has then a normal distribution with mean $\alpha = \alpha_1 + \alpha_2$ and variance $\sigma^2 = \sigma_1^2 + \sigma_2^2$.

This result can be extended to finite sums of independently and normally distributed random variables.

Corollary to Theorem 6.2.1

Let $X_1, \ldots, X_n$ be n independently and normally distributed random variables and write $\alpha_j = \mathscr{E}(X_j)$, $\sigma_j^2 = \text{Var}(X_j)$ for $j = 1, \ldots, n$. The sum $X = \sum_{j=1}^{n} X_j$ is also normally distributed with mean $\alpha = \sum_{j=1}^{n} \alpha_j$ and variance $\sigma^2 = \sum_{j=1}^{n} \sigma_j^2$.

The proof of the corollary is carried by induction.

6.3 The Chi-Square Distribution

In this section, we treat a problem that is often important in statistics. This problem is the determination of the distribution function of the sum of squares of independent normal random variables.

Theorem 6.3.1

Let $X_1, X_2, \ldots, X_n$ be n independently and identically distributed random variables and suppose that the common distribution function of the X_j is normal and has zero mean and variance one. Let $\chi^2 = \sum_{j=1}^{n} X_j^2$, the frequency function of the random variable χ^2 is

$$k_n(x) = \begin{cases} 0 & \text{if } x < 0, \\ \dfrac{1}{2^{n/2}\Gamma\left(\dfrac{n}{2}\right)} x^{(n/2)-1} e^{-x/2} & \text{if } x > 0. \end{cases} \quad (6.3.1)$$

The distribution function corresponding to the frequency (6.3.1) is called the chi-square distribution with n degrees of freedom. We deviate here from our custom to assign capital letters of the Latin alphabet to random variables. This is done in order to be in agreement with the universally accepted notation.

To derive the frequency function (6.3.1) we determine first the frequency function of the square of a normal random variable with zero mean and variance one.

Lemma 6.3.1

Let X be a random variable which has a normal distribution with zero mean and variance one. The frequency function of the random variable X^2 is

$$p(x) = \begin{cases} \dfrac{1}{\sqrt{2\pi}} x^{-1/2} e^{-x/2} & \text{if } x > 0, \\ 0 & \text{if } x < 0. \end{cases}$$

Proof

We first determine the distribution function of the random variable $Y = X^2$. Since Y is nonnegative we have

$$F_Y(x) = 0 \quad \text{for} \quad x < 0, \quad (6.3.2)$$

6.3 THE CHI-SQUARE DISTRIBUTION

while for $x > 0$

$$F_Y(x) = P(Y \le x) = P(X^2 \le x) = P(-\sqrt{x} \le X \le \sqrt{x}). \quad (6.3.3)$$

The frequency function of X is $p_X(x) = \dfrac{1}{\sqrt{2\pi}} \exp\left(-\dfrac{x^2}{2}\right)$, it follows from (6.3.3) that

$$F_Y(x) = \frac{1}{\sqrt{2\pi}} \int_{-\sqrt{x}}^{\sqrt{x}} \exp\left(-\frac{y^2}{2}\right) dy = \frac{2}{\sqrt{2\pi}} \int_0^{\sqrt{x}} \exp\left(-\frac{y^2}{2}\right) dy.$$

We introduce the new variable $u = \dfrac{y^2}{2}$ in the last integral and obtain

$$F_Y(x) = \frac{1}{\sqrt{\pi}} \int_0^{x/2} u^{-1/2} e^{-u} \, du.$$

We differentiate the last equation with respect to x and obtain

$$F'_Y(x) = p_Y(x) = \frac{1}{\sqrt{2\pi}} x^{-1/2} e^{-x/2} = p(x) \quad (x > 0). \quad (6.3.4)$$

We conclude from (6.3.2) and (6.3.4) that

$$F'_Y(x) = p(x) = \begin{cases} 0 & \text{for } x < 0, \\ \dfrac{1}{\sqrt{2\pi}} x^{-1/2} e^{-x/2} & \text{for } x > 0. \end{cases} \quad (6.3.5)$$

Remembering that $\Gamma(\tfrac{1}{2}) = \sqrt{\pi}$ (see Appendix B) we note that $p(x) = k_1(x)$.

We determine next the frequency function of the chi-square distribution with 2 degrees of freedom. The chi-square distribution with 2 degrees of freedom is, by definition, the distribution of the sum of two independent random variables each of which is the square of a random variable having a normal distribution with mean zero and variance one. We see therefore from Theorem 6.1.3 that

$$k_2(x) = \int_{-\infty}^{\infty} k_1(x - y) k_1(y) \, dy. \quad (6.3.6)$$

6 SOME IMPORTANT DISTRIBUTIONS

We substitute (6.3.5) into (6.3.6) and see that

$$k_2(x) = \frac{1}{2\pi} e^{-x/2} \int_0^x \frac{dy}{\sqrt{xy - y^2}}. \tag{6.3.7}$$

It is easily verified by differentiation that

$$\frac{d}{dy}\left[\arcsin\left(\frac{2y}{x} - 1\right)\right] = \frac{1}{\sqrt{xy - y^2}}$$

hence

$$\int_0^x \frac{dy}{\sqrt{xy - y^2}} = \left.\arcsin\left(\frac{2y}{x} - 1\right)\right|_0^x = \arcsin 1 - \arcsin(-1) = \pi.$$

We substitute the value of the integral into (6.3.7) and obtain

$$k_2(x) = \tfrac{1}{2} e^{-x/2}. \tag{6.3.8}$$

This agrees again with formula (6.3.1) of Theorem 6.3.1.

We are now ready to prove Theorem 6.3.1.

It is convenient to carry the proof separately for even and for odd degrees of freedom. For $n = 2m$ formula (6.3.1) becomes

$$k_{2m}(x) = \begin{cases} 0 & \text{for } x < 0, \\ C_{2m} x^{m-1} e^{-x/2} & \text{for } x > 0, \end{cases} \tag{6.3.9a}$$

where $C_{2m} = \dfrac{1}{2^m \Gamma(m)}$.

We see from (6.3.8) that (6.3.9a) is valid for $m = 1$; we prove (6.3.9a) by induction and assume that it is valid for $m = 1, 2, \ldots, n$. We must show that (6.3.9a) holds then also for $m = n + 1$.

It follows from the definition of χ^2 and from Theorem 6.1.3 that

$$k_{2n+2}(x) = \int_{-\infty}^{\infty} k_2(x - y) k_{2n}(y) \, dy. \tag{6.3.10}$$

Since (6.3.9a) is valid for $m = 1$ and was assumed to be valid for $m = n$ we obtain from (6.3.10) for $x > 0$

$$k_{2n+2}(x) = C_2 C_{2n} e^{-x/2} \int_0^x y^{n-1} \, dy = \frac{C_2 C_{2n}}{n} x^n e^{-x/2} \tag{6.3.11}$$

while $k_{2n+2}(x) = 0$ for $x < 0$.

6.3 THE CHI-SQUARE DISTRIBUTION

Since

$$\frac{C_2 C_{2n}}{n} = \frac{1}{2^{n+1} n \Gamma(n)} = \frac{1}{2^{n+1} \Gamma(n+1)} = C_{2n+1}$$

one sees from (6.3.11) that (6.3.9a) is also valid for $m = n + 1$.

We consider next the case of an odd number, $n = 2m - 1$, of degrees of freedom. In this case, (6.3.1) becomes

$$k_{2n-1}(x) = \begin{cases} 0 & \text{for } x < 0, \\ C_{2n-1} x^{[(2n-1)/2]-1} e^{-x/2} & \text{for } x > 0, \end{cases} \quad (6.3.9b)$$

where

$$C_{2n-1} = \frac{1}{2^{(2n-1)/2} \Gamma\left(\frac{2n-1}{2}\right)}.$$

We see from Lemma 6.3.1 that (6.3.9b) is valid for $n = 1$ and prove it again by induction. We again use Theorem 6.1.3 and obtain

$$k_{2n+1}(x) = C_2 C_{2n-1} e^{-x/2} \int_0^x y^{[(2n-1)/2]-1} \, dy$$

$$= \frac{2 C_2 C_{2n-1}}{2n - 1} x^{(2n-1)/2} e^{-x/2}$$

for $x > 0$ while $k_{2n-1}(x) = 0$ for $x < 0$. It is easy to show that

$$\frac{2 C_2 C_{2n-1}}{2n - 1} = \frac{C_{2n-1}}{2n - 1} = C_{2n+1}.$$

Therefore (6.3.9b) holds also for $2n + 1$. This completes the proof of Theorem 6.3.1.

Next we determine the mean and the variance of the chi-square distribution.

Theorem 6.3.2

Let X be a random variable that has a chi-square distribution with n degrees of freedom. Then $\mathscr{E}(X) = n$ and $\text{Var}(X) = 2n$.

We have

$$\mathscr{E}(X) = \frac{1}{2^{n/2}\Gamma\left(\frac{n}{2}\right)} \int_0^\infty x^{n/2} e^{-x/2}\, dx.$$

We introduce the new variable $y = \frac{x}{2}$ in the integral and see that

$$\mathscr{E}(X) = \frac{2\Gamma\left(\frac{n}{2} + 1\right)}{\Gamma\left(\frac{n}{2}\right)} = n. \tag{6.3.12}$$

Similarly

$$\mathscr{E}(X^2) = \frac{1}{2^{n/2}\Gamma\left(\frac{n}{2}\right)} \int_0^\infty x^{(n/2)+1} e^{-x/2}\, dx = \frac{4\Gamma\left(\frac{n}{2} + 2\right)}{\Gamma\left(\frac{n}{2}\right)} = n(n+2).$$

$$\tag{6.3.13}$$

In deriving these expressions we used formula (B.2) of Appendix B. Finally we see from (6.3.12) and (6.3.13) that

$$\operatorname{Var}(X) = \mathscr{E}(X^2) - [\mathscr{E}(X)]^2 = 2n.$$

Tables of the chi-square distribution are available (see Table II). They list values χ_p^2 for which

$$P(\chi^2 > \chi_p^2) = \int_{\chi_p^2}^\infty k_n(x)\, dx = \frac{p}{100}.$$

The value χ_p^2 depends on p and the number n of degrees of freedom; χ_p^2 is called the $p\%$ value of the chi-square distribution with n degrees of freedom.

We conclude this section by mentioning an absolutely continuous distribution, which is of some theoretical as well as practical interest.

6.4 STUDENT'S DISTRIBUTION

Let θ and λ be two positive numbers, it is then easily seen that

$$\int_0^\infty x^{\lambda-1} e^{-\theta x}\, dx = \frac{\Gamma(\lambda)}{\theta^\lambda}.$$

The function

$$p(x) = \begin{cases} \dfrac{\theta^\lambda}{\Gamma(\lambda)} x^{\lambda-1} e^{-\theta x} & \text{for } x > 0, \\ 0 & \text{for } x < 0, \end{cases} \qquad (6.3.14)$$

is therefore a frequency function. The corresponding distribution function is called the gamma distribution.

The chi-square distributions is a particular case of the gamma distribution; we obtain it by putting $\theta = \frac{1}{2}$, $\lambda = \frac{n}{2}$ (n an integer). We also note that (6.3.14) yields for $\lambda = 1$ the frequency function of the exponential distribution, which we introduced in Section 3.4.

6.4 Student's Distribution

The distribution that we shall discuss in this section is very important in mathematical statistics. It was introduced by the British statistician W. S. Gosset who wrote under the pseudonym *Student* and is therefore called Student's distribution. One often refers to it as the *t*-distribution since the letter *t* is used to denote a random variable which has Student's distribution.

Theorem 6.4.1

Let X and Y be two independent random variables and suppose that X has a normal distribution with mean zero and variance one, while Y has a chi-square distribution with n degrees of freedom. Then the random variable

$$t = \frac{X}{\sqrt{Y}} \sqrt{n}$$

has the frequency function

$$s_n(x) = C_n\left(1 + \frac{x^2}{n}\right)^{-(n+1)/2},$$

where

$$C_n = \frac{1}{\sqrt{n\pi}} \frac{\Gamma\left(\frac{n+1}{2}\right)}{\Gamma\left(\frac{n}{2}\right)}.$$

The distribution function $S_n(x)$ whose frequency function is $s_n(x)$ is called Student's distribution (or the t-distribution) with n degrees of freedom.

The random variables X and Y are independent, their joint frequency function is therefore

$$p(x, y) = \frac{1}{\sqrt{2\pi}} \exp\left(-\frac{x^2}{2}\right) \frac{1}{2^{n/2}\Gamma(n/2)} y^{(n/2)-1} e^{-y/2},$$

or

$$p(x, y) = K_n y^{(n/2)-1} \exp\left(-\frac{x^2}{2} - \frac{y}{2}\right), \qquad (6.4.1)$$

where

$$K_n = \frac{1}{2^{n/2}\Gamma\left(\frac{n}{2}\right)\sqrt{2\pi}}. \qquad (6.4.1a)$$

Let $S_n(z)$ be the distribution function of the random variable t. Then

$$S_n(z) = P(t \leq z) = P\left(X \leq \frac{z}{\sqrt{n}}\sqrt{Y}\right).$$

We apply Theorem 6.1.1 and see that

$$S_n(z) = K_n \int_0^\infty y^{(n/2)-1} e^{-y/2} \left[\int_{-\infty}^{z\sqrt{y}/\sqrt{n}} \exp\left(-\frac{x^2}{2}\right) dx\right] dy.$$

6.4 STUDENT'S DISTRIBUTION

We introduce in the inner integral the new variable $x = \dfrac{u\sqrt{y}}{\sqrt{n}}$ and obtain

$$S_n(z) = \frac{K_n}{\sqrt{n}} \int_0^\infty y^{(n-1)/2} e^{-y/2} \left[\int_{-\infty}^z \exp\left(-\frac{u^2 y}{2n}\right) du \right] dy$$

$$= \frac{K_n}{\sqrt{n}} \int_{-\infty}^z \left\{ \int_0^\infty y^{(n-1)/2} \exp\left[-\frac{1}{2}\left(1 + \frac{u^2}{n}\right) y\right] dy \right\} du.$$

We compute the inner integral by introducing the new variable $v = \dfrac{1}{2}\left(1 + \dfrac{u^2}{n}\right) y$, then

$$\int_0^\infty y^{(n-1)/2} \exp\left[-\frac{1}{2}\left(1 + \frac{u^2}{n}\right) y\right] dy = \frac{2^{(n+1)/2}}{\left(1 + \dfrac{u^2}{n}\right)^{(n+1)/2}} \int_0^\infty v^{(n-1)/2} e^{-v}\, dv$$

$$= \frac{2^{(n+1)/2}\, \Gamma\!\left(\dfrac{n+1}{2}\right)}{\left(1 + \dfrac{u^2}{n}\right)^{(n+1)/2}}.$$

Therefore

$$S_n(z) = \frac{\Gamma\!\left(\dfrac{n+1}{2}\right)}{\Gamma\!\left(\dfrac{n}{2}\right)\sqrt{\pi n}} \int_{-\infty}^z \frac{du}{\left(1 + \dfrac{u^2}{n}\right)^{(n+1)/2}}. \qquad (6.4.2)$$

We differentiate (6.4.2) with respect to z and obtain the frequency function of Student's distribution

$$s_n(z) = \frac{\Gamma\!\left(\dfrac{n+1}{2}\right)}{\sqrt{n\pi}\,\Gamma\!\left(\dfrac{n}{2}\right)} \left(1 + \frac{z^2}{n}\right)^{-(n+1)/2}. \qquad (6.4.3)$$

The positive integer n is again called number of the degrees of freedom of Student's distribution.

We note that we obtain the Cauchy distribution if we put $n = 1$ in (6.4.3), the Cauchy distribution is therefore Student's distribution with 1 degree of freedom.

Tables of Student's distribution are available (see Table III). They list values t_p for which $P(|t| > t_p) = 1 - \int_{-t_p}^{t_p} s_n(x)dx = \frac{p}{100}$. The value t_p depends on p and the number n of degrees of freedom, t_p is called a $p\%$ value of Student's distribution with n degrees of freedom.

6.5 Problems

1. Let X and Y be two independently and identically distributed random variables and assume that their common distribution is rectangular over the interval $(0, 1)$. Find the frequency function of $X + Y$.
2. Let X be a random variable that has a chi-square distribution with n degrees of freedom. Let Y be a random variable that has a chi-square distribution with m degrees of freedom. Suppose that X and Y are independent and find the frequency function of $X + Y$.
3. Show that Student's distribution is symmetric.
4. Suppose that the random variable X has a Student's distribution with n degrees of freedom. Determine the frequency function of $Y = |X|$.
5. Suppose that the random variable X has Student's distribution with 10 degrees of freedom. Find a value x such that

    (a) $P(X \leq x) = 0.99$;

    (b) $P(|X| \leq x) = 0.99$.

6. Suppose that the random variable X has a chi-square distribution with 15 degrees of freedom. Find the value x such that $P(X^2 \leq x) = 0.95$.
7. Let X be a random variable that has Student's distribution with n degrees of freedom. Determine the frequency function of X^2.

6.5 PROBLEMS

8. Let X be a random variable which has a Cauchy distribution. Find the frequency function of X^2.

9. Let X be a random variable that has a chi-square distribution. Find the frequency function of the random variable $Y = \dfrac{X}{1+X}$.

10. Let X and Y be two independent discrete random variables that can assume only nonnegative integer values and let $p_j = P(X=j)$, $q_k = P(Y=k)$. Show that the probabilities of the sum $X+Y$ are given by $P(X+Y=m) = \sum_{l=0}^{m} p_{m-l} q_l$.

11. Let X and Y be two independent Poissonian random variables with $\mathscr{E}(X) = \mathscr{E}(Y) = \lambda$ and find the distribution of $X+Y$.

12. Let X and Y be two independent Poissonian variables and assume $\mathscr{E}(X) = \lambda$; $\mathscr{E}(Y) = \mu$. Find the distribution of $X+Y$.

13. Let $X_1, \ldots, X_n$ be n independently and identically distributed random variables and assume that their common distribution is a Poisson distribution with parameter λ. Find the distribution of $S_n = \sum_{j=1}^{n} X_j$.

14. Let X and Y be two independently and identically distributed random variables and suppose that their common distribution function is an exponential distribution with parameter θ. Find the frequency function of the sum $X+Y$.

15. Let X_1, X_2, X_3 be three independently and identically distributed random variables and suppose that their common distribution function is an exponential distribution with parameter θ. Use the result of the preceding problem to show that the frequency function of the sum $X_1 + X_2 + X_3$ is given by $p(x) = \dfrac{\theta^3 x^2 e^{-\theta x}}{2}$ if $x > 0$ and $p(x) = 0$ for $x < 0$.

16. Let $X_1, X_2, \ldots, X_n$ be n independently and identically distributed random variables and suppose that their common distribution function is an exponential distribution with parameter θ. Show that the frequency function of the sum $X_1 + X_2 + \cdots + X_n$ is a gamma

distribution with parameters $\lambda = n$ and θ. [*Hint*: Use mathematical induction.]

17. Let X be a random variable which has Student's distribution with 2 degrees of freedom. Show that the first moment of X exists, and compute the first algebraic moment and the first absolute moment of X. Does the second moment of X exist? Justify your answer.

18. Let X be a random variable that has Student's distribution with 3 degrees of freedom. Show that the mean and variance of X exist and determine the variance of X.

19. Let X be a random variable which has a chi-square distribution with n degrees of freedom. Find the third moment of X.

20. Let X be a random variable that has the frequency function

$$p(x) = \begin{cases} 0 & \text{if } |x| > 1, \\ 1 - |x| & \text{if } |x| \leq 1, \end{cases}$$

and suppose that the random variable Y is independently distributed of X and has a uniform distribution over the interval $(0, 1)$. Find the frequency function and the distribution function of the random variable $Z = X + Y$.

II | MATHEMATICAL STATISTICS

The discussions of Chapter 6 are typical for problems in probability theory: Certain random variables and their distribution functions are completely given, new random variables are defined as functions of the known random variables and the distributions of the new variables are determined.

In this part we shall develop methods of mathematical statistics and we shall study procedures that permit confrontation of the mathematical model with empirical data.

7 Sampling

7.1 Statistical Data

Mathematical statistics is based on probability theory, but has a different character. One considers a phenomenon that shows statistical regularity and constructs a probabilistic model to describe it. The distributions of the random variables occurring in the model are not completely known, but the quantities that are represented in the model by random variables can be observed repeatedly. The statistician uses a number of independent observations of these quantities to arrive at conclusions about the distribution functions of the model.

Let us consider a specific situation by assuming that the model deals with a single random variable X. Suppose that the distribution function $F(x)$ of X contains one or more unknown parameters so that $F(x)$ is not completely determined. However, one can obtain n independent observations of the quantity that is being studied. This means, in the language of probability theory, that one considers n random variables $X_1, \ldots, X_n$ which are independently and identically distributed with common distribution function $F(x)$. The actual observations $x_1, \ldots, x_n$ are then the values assumed by the $X_1, \ldots, X_n$. The $X_1, \ldots, X_n$ are the sample taken from a hypothetical population† of possible observations. The number n of elements in the sample (number of observations) is called the sample size. The set $(x_1, \ldots, x_n)$ of the n actually observed values is called a realization of the sample. The possible values of the $(X_1, \ldots, X_n)$ can be regarded as points in an n-dimensional euclidean space, which is called the sample space. Any realization of the sample is then a point in the sample space. Let $S(x_1, \ldots, x_n)$ be a function of n variables such that $S \in \mathfrak{M}_n$. Then $S(X_1, \ldots, X_n)$ is a random variable and is called a statistic. The distribution function $F(x)$ is called the population distribution function.

As a concrete example, we consider the tossing of a possibly biased coin. Let p be the unknown probability of heads and let X be an indicator random variable which assumes the value 1 when the result of a toss is heads but 0 if it is tails. The random variable X has a discrete distribution

$$F(x \,;\, p) = p\varepsilon(x - 1) + (1 - p)\varepsilon(x).$$

This distribution function contains the unknown parameter p and is therefore not completely known. One performs n independent trials with the coin, that is one considers n independent random variables with common distribution function $F(x \,;\, p)$. This leads to a sequence

---

† The population may be a human or an animal population, or it could be the yield of certain plots, or it may consist of certain items manufactured in a process during a certain time (for instance, radio tubes or resistors produced in a working day). In all these cases some characteristics of the elements of the population are studied by selecting a sample at random from the population and by measuring the characteristic that is under investigation.

$(x_1, \ldots, x_n)$, where x_j is the realization (outcome) of the jth toss, that is, x_j is either 0 or 1. The statistician's task is to use these observed data to estimate the unknown parameter p. As a result of this estimation he can assert that the coin is unbiased if his estimate is reasonably close to $\frac{1}{2}$.

7.2 Sample Characteristics

We consider a population with population distribution function $F(x)$ and take a sample $X_1, \ldots, X_n$ of size n. Let $nF_n^*(x)$ be the number of subscripts k for which $X_k \leq x$; this means that $F_n^*(x)$ is the relative frequency of the elements of the sample that do not exceed x. Therefore

$$F_n^*(x) = \frac{1}{n} \sum_{j=1}^{n} \varepsilon(x - X_j). \qquad (7.2.1)$$

The function $F_n^*(x)$ is a distribution function, it is called the sample distribution function. That is, $F_n^*(x)$ is a discrete distribution with n equal jumps at the points of the sample.

Since $F_n^*(x)$ is a distribution function one can introduce its typical values [see Chapter 4]. The typical values of $F_n^*(x)$ are called sample characteristics.

Since $F_n^*(x)$ has a finite number of discontinuity points, it has finite moments of all orders. Let a_k be the moment of order k of $F_n^*(x)$; clearly

$$a_k = \frac{1}{n} \sum_{j=1}^{n} X_j^k. \qquad (7.2.2)$$

The quantity a_k is called the sample moment of order k; it is a random variable whose possible values are the numbers $\frac{1}{n} \sum_{j=1}^{n} x_j^k$, where $(x_1, \ldots, x_n)$ is a realization of the sample. The sample moment of order one is called the sample mean and is usually written as

$$\overline{X} = \frac{1}{n} \sum_{j=1}^{n} X_j. \qquad (7.2.3)$$

One can also introduce central sample moments,

$$m_k = \frac{1}{n} \sum_{j=1}^{n} (X_j - \overline{X})^k. \qquad (7.2.4)$$

The most important of these is the sample variance, usually denoted by the letter s^2 rather than by m_2,

$$s^2 = \frac{1}{n} \sum_{j=1}^{n} (X_j - \overline{X})^2 = \frac{1}{n} \sum_{j=1}^{n} X_j^2 - \overline{X}^2 = a_2 - a_1^2. \qquad (7.2.5)$$

As a rule, we denote the typical values of the population distribution by lowercase Greek letters and the corresponding sample characteristic by lowercase Latin letters. The notation $\overline{X}$ for the sample mean is an exception, which is justified by the universally accepted usage.

It is also possible to define sample characteristics that correspond to the correlation coefficient and to the regression coefficients.

Let $(X_1, Y_1), (X_2, Y_2), \ldots, (X_n, Y_n)$ be a sample of size n taken from a bivariate population. The sample variances of X and of Y are given by (7.2.5) and are

$$s_X^2 = \frac{1}{n} \sum_{j=1}^{n} (X_j - \overline{X})^2,$$

$$s_Y^2 = \frac{1}{n} \sum_{j=1}^{n} (Y_j - \overline{Y})^2,$$

where $\overline{X}$ and $\overline{Y}$ are the sample means. We define the sample covariance by

$$s_{XY} = \frac{1}{n} \sum_{j=1}^{n} (X_j - \overline{X})(Y_j - \overline{Y}). \qquad (7.2.6)$$

The sample correlation coefficient is then defined to be

$$r = \frac{s_{XY}}{s_X s_Y}. \qquad (7.2.7)$$

It can again be shown that $|r| \leq 1$.

It is also possible to determine regression lines and regression coefficients for the sample by means of the method of least squares.

7.2 SAMPLE CHARACTERISTICS

The sample regression coefficients b_0 and b_1 are numbers, depending on the observations, which minimize the function

$$G(b_0, b_1) = \sum_{j=1}^{n} [Y_j - b_0 - b_1 X_j]^2.$$

It is easily seen that

$$G(b_0, b_1) = ns_Y^2 + n(b_0 - \overline{Y} + b_1 \overline{X})^2 + nb_1^2 s_X^2 - 2nb_1 s_{XY}. \quad (7.2.8)$$

It follows from (7.2.8) that

$$\frac{\partial G}{\partial b_0} = 2n(b_0 - \overline{Y} + b_1 \overline{X}),$$

$$\frac{\partial G}{\partial b_1} = 2n\overline{X}(b_0 - \overline{Y} + b_1 \overline{X}) + 2nb_1 s_X^2 - 2ns_{XY}.$$

The sample regression coefficients satisfy the system of linear equations

$$\begin{aligned} b_0 - \overline{Y} + b_1 \overline{X} &= 0, \\ b_1 s_X^2 - s_{XY} &= 0. \end{aligned} \quad (7.2.9)$$

This system of equations has the solution

$$b_0 = \overline{Y} - \overline{X} \frac{s_{XY}}{s_X^2},$$

$$b_1 = \frac{s_{XY}}{s_X^2}.$$

Using (7.2.7), we obtain for the sample regression coefficients the expressions

$$b_0 = \overline{Y} - \overline{X} \frac{rs_Y}{s_X},$$

$$b_1 = \frac{rs_Y}{s_X}, \quad (7.2.10)$$

which are analogous to the population regression coefficients given by (4.5.10).

The straight line

$$y = b_0 + b_1 x \quad (7.2.11)$$

is called the sample regression of Y on X. In a similar manner one can also define the sample regression line of X on Y.

7.3 Moments and Distributions of Sample Characteristics

Sample characteristics are random variables, it is, therefore, of interest to determine their moments and also their distribution functions. First we compute the moments of some sample characteristics.

Let $X_1, \ldots, X_n$ be a sample from a population. Then

$$\mathscr{E}(\overline{X}) = \frac{1}{n} \sum_{j=1}^{n} \mathscr{E}(X_j). \tag{7.3.1}$$

Since the elements of a sample are identically distributed random variables we have

$$\mathscr{E}(\overline{X}) = \alpha, \tag{7.3.2}$$

where α is the population mean. Similarly we see from (7.3.1) that†

$$\mathrm{Var}(\overline{X}) = \frac{1}{n^2} \sum_{j=1}^{n} \mathrm{Var}(X_j) = \frac{1}{n} \sigma^2. \tag{7.3.3}$$

One can also determine higher moments of sample characteristics. As an example we determine the third moment of $\overline{X}$. It is easily seen that

$$\left(\sum_{j=1}^{n} X_j \right)^3 = \sum_{j=1}^{n} X_j^3 + 3 \sum_{\substack{j=1 \\ j \neq k}}^{n} \sum_{k=1}^{n} X_j^2 X_k + 6 \sum_{\substack{j=1 \\ j \neq k, j \neq l \\ k \neq l}}^{n} \sum_{k=1}^{n} \sum_{l=1}^{n} X_j X_k X_l.$$

Therefore,

$$\mathscr{E}\left[\left(\sum_{j=1}^{n} X_j \right)^3 \right] = n\alpha_3 + 3n(n-1)\alpha_2 \alpha_1 + n(n-1)(n-2)\alpha_1^3,$$

---

† Here we also used the fact that the elements of a sample are by definition independent random variables.

7.3 MOMENTS AND DISTRIBUTIONS

so that

$$\mathscr{E}(\bar{X}^3) = \frac{\alpha_3 + 3(n-1)\alpha_2\alpha_1 + (n-1)(n-2)\alpha_1^3}{n^2}. \qquad (7.3.4)$$

In deriving (7.3.1)–(7.3.4), it was tacitly assumed that the moments on the right-hand sides of these equations exist.

Next we determine the mean and the variance of the sample variance s^2.

We see from (7.2.5) that

$$\mathscr{E}(s^2) = \mathscr{E}(a_2 - a_1^2). \qquad (7.3.5)$$

One has

$$\mathscr{E}(a_2) = \mathscr{E}\left(\frac{1}{n}\sum_{j=1}^n X_j^2\right) = \alpha_2, \qquad (7.3.6a)$$

while

$$\mathscr{E}(a_1^2) = \frac{1}{n^2}\mathscr{E}\left[\left(\sum_{j=1}^n X_j\right)^2\right]$$

$$= \frac{1}{n^2}\mathscr{E}\left[\sum_{j=1}^n X_j^2 + \sum_{\substack{j=1 \\ j\neq k}}^n \sum_{k=1}^n X_j X_k\right],$$

so that

$$\mathscr{E}(a_1^2) = \frac{\alpha_2 + (n-1)\alpha_1^2}{n}. \qquad (7.3.6b)$$

It follows from (7.3.5), (7.3.6a), and (7.3.6b) that

$$\mathscr{E}(s^2) = \frac{n-1}{n}(\alpha_2 - \alpha_1^2) = \frac{n-1}{n}\sigma^2. \qquad (7.3.7)$$

Since $\operatorname{Var}(s^2) = \mathscr{E}(s^4) - [\mathscr{E}(s^2)]^2$ one must find $\mathscr{E}(s^4)$. We have

$$s^4 = (a_2 - a_1^2)^2 = a_2^2 - 2a_2 a_1^2 + a_1^4.$$

A somewhat tedious but elementary computation† yields

$$\operatorname{Var}(s^2) = \frac{\mu_4 - \mu_2^2}{n} - \frac{2(\mu_4 - 2\mu_2^2)}{n^2} + \frac{\mu_4 - 3\mu_2^2}{n^3}. \qquad (7.3.8)$$

It is not difficult to compute higher moments, the computations are, however, very tedious.

The expected value and the variance of a sample characteristic give some valuable information about its behavior. Complete insight can be obtained if the distribution function of the sample characteristic is known. It is usually difficult to determine the distribution function of a statistic, we shall discuss here only some simple but important cases.

Theorem 7.3.1

Let $X_1, \ldots, X_n$ be a sample from a normal population and suppose that the population distribution function has mean α and variance σ^2. The sample mean $\overline{X} = \dfrac{1}{n}\sum_{j=1}^{n} X_j$ is then also normally distributed with mean α and standard deviation $\dfrac{1}{\sqrt{n}}\sigma$.

We see from Theorem 6.2.1 that the sum $\sum_{j=1}^{n} X_j$ is normally distributed with mean $n\alpha$ and variance $n\sigma^2$. Let $F_{\overline{X}}(x)$ be the distribution function of $\overline{X}$, then

$$\begin{aligned} F_{\overline{X}}(x) &= P\left(\frac{1}{n}\sum_{j=1}^{n} X_j \le x\right) \\ &= P\left(\sum_{j=1}^{n} X_j \le nx\right) \\ &= \frac{1}{\sigma\sqrt{2\pi n}} \int_{-\infty}^{nx} \exp\left[-\frac{1}{2n\sigma^2}(y - n\alpha)^2\right] dy, \end{aligned}$$

---

† See [*1*, p. 348].

7.3 MOMENTS AND DISTRIBUTIONS

so that

$$F'_{\bar{X}}(x) = \frac{\sqrt{n}}{\sigma\sqrt{2\pi}} \exp\left[-\frac{n}{2\sigma^2}(x-\alpha)^2\right],$$

and the statement is proven.

Theorem 7.3.1 indicates that the probability distribution of $\bar{X}$ shows a greater concentration about the mean than the population distribution function. While this is a frequent occurrence, exceptions do exist.

Let X_1 and X_2 be two independently and identically distributed random variables each having a Cauchy distribution with frequency function $p(x) = \dfrac{1}{\pi(1+x^2)}$. Let $Z = X_1 + X_2$; the frequency function of Z is then

$$p_Z(x) = \frac{1}{\pi^2} \int_{-\infty}^{\infty} \frac{dy}{[1+(x-y)^2](1+y^2)}.$$

This is the integral of a rational function and can be evaluated by the usual methods. One obtains

$$p_Z(x) = \frac{2}{\pi(x^2+4)}.$$

Let $Y = \tfrac{1}{2}(X_1 + X_2) = \tfrac{1}{2}Z$; the frequency function of Y, is then

$$p_Y(x) = \frac{1}{\pi(1+x^2)}.$$

The distribution function of Y is, therefore, identical with the common distribution of X_1 and X_2. This result can be generalized by elementary but tedious computations† and one obtains the following theorem.

---

† Methods that are more advanced than the ones used in this book permit an easy derivation of Theorem 7.3.2 (see Cramér [1, p. 247]).

Theorem 7.3.2

Let $X_1, \ldots, X_n$ be a sample from a population that has a Cauchy distribution, the sample mean $\bar{X}$ has then the same distribution as every X_j.

It is often difficult to derive the exact distribution of sampling characteristics (or other statistics), but one can obtain useful approximations by using the central limit theorem.

We see from Theorem 5.2.3 that $\bar{X}$ is asymptotically normally distributed† with mean α and variance $\dfrac{\sigma^2}{n}$ for all populations whose population distribution has a finite second moment.

It can also be shown‡ that the sample moments a_k are also asymptotically normally distributed with mean α_k and variance $\dfrac{\alpha_{2k} - \alpha_k^2}{n}$.

However, not all sample characteristics have an asymptotically normal distribution. It can be shown that the distribution of the smallest or of the largest value in the sample does in general not approximate the normal distribution, no matter how large the sample size.§

We finally state two very important theorems that require for their proofs certain results from advanced calculus and from the theory of linear transformations. We are therefore not able to prove these theorems here, but we shall discuss one of their applications.

Theorem 7.3.3

Let $X_1, \ldots, X_n$ be a sample from a population whose distribution function is normal¶ with mean α and variance σ^2. Let $\bar{X} = \dfrac{1}{n} \sum\limits_{j=1}^{n} X_j$ be

---

† This means that the distribution function of $\bar{X}$ can be approximated by a normal distribution if n is sufficiently large.

‡ See Cramér [*1*, p. 364].

§ See Cramér [*1*, p. 370].

¶ For the sake of brevity, we express this by stating alternatively: Let $X_1, \ldots, X_n$ be a sample from a normal population with mean α and variance σ^2.

7.3 MOMENTS AND DISTRIBUTIONS

the sample mean and let $s^2 = \frac{1}{n}\sum_{j=1}^{n}(X_j - \overline{X})^2$ be the sample variance. Then the random variables $\overline{X}$ and s^2 are independent.

Theorem 7.3.4†

Let $X_1, \ldots, X_n$ be a sample from a population whose population distribution function is normal with mean α and variance σ^2. Let s^2 be the sample variance, then $\frac{ns^2}{\sigma^2}$ has a chi-square distribution with $n-1$ degrees of freedom.

Remark

Theorem 7.3.3 indicates that we must distinguish functional independence from stochastic independence (that is, independence in the sense of probability theory). The sample mean and the sample variance are certainly functionally dependent, since they contain the same variables $X_1, \ldots, X_n$, but this is compatible with their stochastic independence as stated in Theorem 7.3.3.

We consider next an application of the last two theorems.

Let $X_1, \ldots, X_n$ be a sample from a population whose distribution function is normal with mean α and variance σ^2. It follows from Theorem 7.3.1 that the random variable

$$Y = \sqrt{n}\frac{\overline{X} - \alpha}{\sigma} \qquad (7.3.9)$$

has a standardized normal distribution. Let

$$Z = \frac{ns^2}{\sigma^2}, \qquad (7.3.10)$$

the random variable Z, has, by Theorem 7.3.4, a chi-square distribution with $n-1$ degrees of freedom. We see, moreover, from Theorem 7.3.3 that Y and Z are independent; we can, therefore, apply Theorem 6.4.1 and obtain the following result.

---

† For the proofs of Theorems 7.3.3 and 7.3.4 see [1, p. 382] or [2, p. 159]

Theorem 7.3.5

Let $X_1, \ldots, X_n$ be a sample from a normal population with mean α and variance σ^2. The statistic $t = \dfrac{\sqrt{n-1}(\overline{X} - \alpha)}{s}$ has then Student's distribution with $n-1$ degrees of freedom.

Remark

The expression of t does not depend on σ, and the distribution of the t statistic depends only on the degrees of freedom, not on the parameters α and σ^2 of the population.

The sampling distributions of characteristics† of certain multivariate samples are also known. Several authors studied sampling from a bivariate normal population and derived the distribution of the sample correlation coefficient r. This distribution is comparatively simple if the population correlation coefficient $\rho = 0$. It can be shown that, in this case,

$$t = \sqrt{n-2}\,\frac{r}{\sqrt{1-r^2}} \tag{7.3.11}$$

has Student's distribution with $n-2$ degrees of freedom.

The sampling distribution of regression coefficients is also known, it can be shown that

$$t = \frac{s_X \sqrt{n-2}}{s_Y \sqrt{1-r^2}} (b_{21} - \beta_{21}) \tag{7.3.12}$$

has Student's distribution with $n-2$ degrees of freedom.

A detailed discussion of these results would exceed the scope of this book, but can be found in [*1*, p. 397 ff].

---

† See [*1*, p. 400 ff].

7.4 Problems

1. Find the sampling distribution of the sample mean when the population has a Poisson distribution with expected value λ. [*Hint*: Use the result of Problem 13 of Chapter 6.]
2. Determine the first and second moments and the variance of the sampling distribution of Problem 1.
3. Two sets of observations of the same variable are taken. Let $\bar{X}_1$ and $\bar{X}_2$ be their sample means and n_1 and n_2 the corresponding sample sizes. The two samples are combined, what is the sample mean of the combined sample of $n_1 + n_2$ observations?
4. Two samples, one comprising n_1 the other n_2 observations of the same variable are taken. The sample means $\bar{X}_1$ and $\bar{X}_2$ as well as the sample variances s_1^2 and s_2^2 of the two samples are obtained. The two samples are combined; what is the sample variance of the combined sample of $n_1 + n_2$ observations?
5. Yeast cells were counted in the 400 squares of a hemacytometer. The data given in Table 7.1 were obtained. Compute the mean, the

Table 7.1

Cells counted	Number of squares
0	213
1	128
2	37
3	18
4	3
5	1

variance and the third moment of this sample. (Data taken from [4, p. 351].)
6. Let $X_1, \ldots, X_n$ be a sample of size n from a population whose distribution function is rectangular over the interval $(-1, +1)$. Find the mean and the variance of its sample mean.

7. Let $X_1 \cdots X_n$ be a sample of size n from a population whose distribution is rectangular over the interval (a, b). Find the mean and the variance of its sample mean.

8. Let $X_1 \cdots X_n$ be a sample from a population whose distribution is given by the frequence function

$$p(x) = \begin{cases} |x| & \text{if } |x| < 1, \\ 0 & \text{otherwise.} \end{cases}$$

Find the mean and the variance of its sample mean.

9.† Two methods were used in a study of the latent heat of fusion of ice. Both Method A (an electrical method) and Method B (a method of mixtures) were conducted with the specimens cooled to $-0.72°C$. The data represent the change in total heat from $-0.72°C$ to water at $0°C$, in calories per gram of mass (see Table 7.2).

Table 7.2

Change in Total Heat from $-0.72°C$ to $0°C$[a]

Method A	Method B
79.98	80.02
80.04	79.94
80.02	79.98
80.04	79.97
80.03	80.03
80.03	79.95
80.04	79.97
79.97	
80.05	
80.03	
80.02	
80.00	
80.02	

[a] Values given in calories per gram mass.

Compute the sample mean and the sample variance for each of the two samples.

† Taken from [3, p. 3.23].

10.† An investigator determined the compressive strength of concrete and obtained the following measurements (in pounds per square inch) for his sample:

$$1939$$
$$1697$$
$$3030$$
$$2424$$
$$2020$$
$$2909$$
$$1815$$
$$2020$$
$$2310$$

Determine the mean and variance of this sample.

11. Let $X_1, X_2, \ldots, X_m$ be a sample of size m taken from a population whose distribution function is a chi-square distribution with n degrees of freedom. Let $\bar{X}$ be the sample mean and determine the mean and the variance of the distribution of $\bar{X}$.

12. Let $X_1, X_2, \ldots, X_n$ be a sample of size n taken from a population whose distribution function is an exponential distribution. Let $\bar{X}$ be the sample mean and determine its mean and its variance.

13. Let $X_1, X_2, \ldots, X_n$ be a sample of size n taken from a population whose distribution function is a Cauchy distribution and let $\bar{X}$ be the sample mean. What is the mean of the sampling distribution of $\bar{X}$?

14. Let $X_1, X_2, \ldots, X_n$ be a sample of size n taken from a population whose distribution is a gamma distribution. Determine the mean and the variance of the sample mean. [*Hint:* Use the information concerning the gamma function contained in Appendix B.]

References

1. H. Cramér, "Mathematical Methods of Statistics," Princeton Univ. Press, Princeton, New Jersey, 1946.

---

† Taken from [*3*, p. 3.26].

2. H. Tucker, "An Introduction to Probability Theory and Mathematical Statistics," Academic Press, New York, 1962.
3. M. Natrella "Experimental Statistics (NBS Handbook 99)," U.S. Govt. Printing Office, Washington, D.C., 1963.
4. "Student," On the error of counting by a haemacytometer. *Biometrika* **V**, 351 (1907); "Students' Collected Papers," Cambridge Univ. Press, London and New York, 1958.

8 | Estimation

The statistician has some incomplete information concerning a population. This information is usually provided by a sample and should be used to gain some additional knowledge of the population. In many cases, some *a priori* information concerning the population is also available. For instance, it is often justified to assume that the population distribution function has a given form, but contains some unknown parameters.† In such a situation, the statistician must use the sample to obtain an estimate of the unknown parameters.

---

† For example, we could know that the population distribution function is normal with unknown mean but known variance or it could be normal with both mean and variance unknown.

First we consider the case where the functional form of the population distribution function is known, but contains only one unknown parameter. Let $F(x|\theta)$ be the population distribution function where θ is the unknown parameter. A sample $X_1, \ldots, X_n$ of size n is drawn from the population and it is desired to find an estimate for the unknown constant θ. This can be accomplished by constructing a function $\theta^* = \theta^*(X_1, \ldots, X_n)$ of the sample and by using θ^* as an estimate for θ. The choice of the function θ^* is not quite arbitrary, it must be a statistic (that is, an element of $\mathfrak{M}_n$) that should satisfy certain requirements. These will justify its use as an estimate for the parameter θ. The function θ^* is called a point estimate of θ.

Alternatively one could construct two statistics θ_1^* and θ_2^* of the sample in such a way that $\theta_1^* < \theta_2^*$ for all $X_1, \ldots, X_n$ and that the probability that the interval (θ_1^*, θ_2^*) covers the unknown parameter θ has a preassigned value. This method of estimation is called estimation by interval. In Section 8.2 we discuss point estimates. Section 8.3 deals with interval estimation.

8.1 Properties of Estimates

The estimate $\theta^* = \theta^*(X_1, \ldots, X_n)$ is a random variable, its distribution will, in general, also depend on the unknown parameter θ. A function θ^* of the sample will be a good estimate if it is—in some sense—close to the unknown θ. This is of course a very vague formulation, the purpose of this section is the listing of some properties which characterize good estimates.

An estimate $\theta^* = \theta^*(X_1, \ldots, X_n)$ is said to be unbiased if the expectation $\mathscr{E}(\theta^*)$ exists and if $\mathscr{E}(\theta^*) = \theta$, no matter what the value of θ is.

The difference $b(\theta^*) = \mathscr{E}(\theta^*) - \theta$ is called the bias of the estimate θ^*.

Suppose that the parameter to be estimated is the population mean. We see from formula (7.3.2) that the sample mean is always an unbiased estimate of the population mean for any population.

8.1 PROPERTIES OF ESTIMATES

We had [formula (7.3.7)] $\mathscr{E}(s^2) = \dfrac{n-1}{n} \sigma^2$, the sample variance is, therefore, not an unbiased estimate of the population variance. The bias of the sample variance is $b(s^2) = -\dfrac{\sigma^2}{n}$. It is easy to obtain an unbiased estimate for the population variance, it follows from (7.3.7) that

$$\frac{n}{n-1} s^2 = \frac{1}{n-1} \sum_{j=1}^{n} (X_j - \overline{X})^2$$

is an unbiased estimate of σ^2.

The property of unbiasedness refers to a fixed sample size and to a single estimate. It is often desirable to study sequences of estimates $\theta_n^* = \theta_n^*(X_1, \ldots, X_n)$, where the estimate θ_n^* is obtained from a sample of size n. In the preceding pages, we considered the sample mean and the sample variance for a fixed sample size, if we let n run through all positive integers, we see that $\overline{X}$ and s^2 can also be considered as sequences of estimates. We shall write $\overline{X}_n$ and s_n^2 instead of $\overline{X}$ and s^2 whenever we emphasize the dependence of these statistics on the sample size and we wish to consider sequences of these estimates for the sample mean and for the sample variance.

A sequence θ_n^* of estimates is said to be a consistent sequence of estimators for a parameter θ if $\lim\limits_{n \to \infty} P(|\theta_n^* - \theta| > \varepsilon) = 0$ for every $\varepsilon > 0$, irrespective of the value of θ.

Consistency of a sequence of estimates means that the probability mass of the distribution function of θ_n^* is concentrated in an arbitrarily small interval around the unknown parameter value θ, provided that n is sufficiently large.

Next we consider a population whose population distribution function has finite second moments. It follows from (7.3.3) that $\lim\limits_{n \to \infty} \text{Var}(\overline{X}_n) = 0$ and we conclude from Chebyshev's inequality that the sequence of sample means is a sequence of consistent estimates for the population mean. More generally, we prove in the same way the following result.

Theorem 8.1.1

Suppose that the population distribution function $F(x|\theta)$ of a population has finite second moments and let $\theta_n^* = \theta_n^*(X_1, \ldots, X_n)$ be a sequence of unbiased estimates for the population parameter θ and assume that $\lim_{n\to\infty} \text{Var}(\theta_n^*) = 0$, then the θ_n^* are a consistent sequence of estimates for θ.

It follows from Theorem 8.1.1 that the sequence $\dfrac{n}{n-1} s_n^2$ is a sequence of consistent estimates for σ^2.

Remark

It can also be shown that the sequence s_n^2 of sample variances (without the correction for bias) is also a sequence of consistent estimates. This indicates that the conditions of Theorem 8.1.1 are sufficient but not necessary.

It can happen that more than one estimate exists for a population parameter. Suppose that there exist two unbiased estimates θ_1^* and θ_2^* of the population parameter θ. Clearly, that estimate whose distribution function is more highly concentrated about θ will be preferable; this means that in the case of unbiased estimates, we should select the estimates with the smaller variance. We say that the estimate θ_1^* is more efficient than θ_2^* if

$$\mathcal{E}[(\theta_1^* - \theta)^2] \leq \mathcal{E}[(\theta_2^* - \theta)^2].$$

The quotient

$$\frac{\mathcal{E}[(\theta_1^* - \theta)^2]}{\mathcal{E}[(\theta_2^* - \theta)^2]}$$

is called the relative efficiency of θ_1^* with respect to θ_2^*.

A similar reasoning can be applied if more than two estimates are available. The "best" estimate will be the one which has the smallest variance. If the population distribution and the estimate satisfy certain conditions it is possible to find a lower bound for the variance of all estimates which satisfy these conditions. It is also possible to decide whether an estimate attains the minimum variance. An estimate which attains the minimum variance is called an efficient estimate. The exact formulation of these conditions as well as the derivation of the lower bound would exceed the scope of this book.† We mention here only that the sample mean of a normal population as well as the sample mean of a Poisson population are efficient estimates.

A somewhat similar asymptotic property of sequences of estimates is often of interest. Let $\theta_n^* = \theta_n^*(X_1, \ldots, X_n)$ be a sequence of estimates for a population parameter θ. The distribution functions of the θ_n^* will often satisfy the conditions of the central limit theorem and will in these cases converge to a normal distribution.

A sequence θ_n^* of estimates is said to be asymptotically efficient if the following two conditions are satisfied.

(i) The distribution function of $\sqrt{n}\,(\theta_n^* - \theta)$ tends to a normal distribution with zero mean and finite variance σ^2 as n increases.

(ii) If T_n^* is a second sequence which satisfies (i) and if τ^2 is the variance of the corresponding limiting distribution then $\dfrac{\sigma^2}{\tau^2} \leq 1$.

8.2 Point Estimation

In this section, we discuss two methods for the construction of estimates. The first of these is due to R. A. Fisher.

Let $X_1, \ldots, X_n$ be a sample of size n taken from a population whose population distribution function is absolutely continuous and depends on exactly one parameter θ. Let $p(x|\theta)$ be the frequency function of the population. We define the likelihood function

$$L(X_1, \ldots, X_n | \theta) = \prod_{j=1}^{n} p(X_j | \theta). \qquad (8.2.1)$$

This is a function of the sample and of the unknown parameter.

---

† A detailed discussion can be found in Cramér [1, p. 477 ff].

The definition has to be modified if the sample comes from a population which has a discrete population distribution function. Let $\{x_i\}$ be the countable set of possible values of the population distribution function and write $p_i(\theta) = P(X = x_i)$. Suppose that the possible value x_i was observed f_i times where $f_i \geq 0$, $\sum_i f_i = n$. We define the likelihood function as

$$L(X_1, \ldots, X_n | \theta) = \prod_i [p_i(\theta)]^{f_i}. \tag{8.2.2}$$

The likelihood function is again a function of the sample and of the unknown paramater θ. The maximum likelihood method uses as an estimate $\hat{\theta}(X_1, \ldots, X_n)$ of θ the value that maximizes the likelihood function.

The determination of the maximum likelihood estimate becomes, therefore, a problem of calculus. Instead of maximizing L, it is convenient to maximize $\log L$ and to solve the likelihood equation

$$\frac{\partial \log L}{\partial \theta} = 0. \tag{8.2.3}$$

One must determine the roots of the likelihood equation and consider those which yield relative maxima. All roots of the likelihood equations are functions of the sample, the root which yields the greatest relative maximum is called the maximum likelihood estimate and is denoted

$$\hat{\theta} = \hat{\theta}(X_1, \ldots, X_n).$$

We mention without proof† the following properties, which indicate the importance of maximum likelihood estimates.

Theorem 8.2.1

If an efficient estimate θ^* of θ exists then the likelihood equation has a unique solution that is equal to θ^*.

---

† For the proof of Theorem 8.2.1 see [*1*, p. 498 ff], for Theorem 8.2.2 see [*1*, p. 500 ff].

8.2 POINT ESTIMATION

Theorem 8.2.2

Under certain, not too restrictive conditions, a sequence $\hat{\theta}_n$ of maximum likelihood estimates is a consistent sequence of estimates and is also an asymptotically efficient sequence of estimates.

The maximum likelihood method can be extended to the case of several unknown parameters. Suppose that the population distribution function depends on k unknown parameters $\theta_1, \ldots, \theta_k$. The likelihood function will then have the form

$$L(X_1, \ldots, X_n | \theta_1, \ldots, \theta_k) = \prod_{j=1}^{n} p(X_j | \theta_1, \ldots, \theta_k) \quad (8.2.4a)$$

in case the population distribution function is absolutely continuous with frequency function $p(x | \theta_1, \ldots, \theta_k)$, but

$$L(X_1, \ldots, X_n | \theta_1, \ldots, \theta_k) = \prod_i [p_i(\theta_1, \ldots, \theta_k)]^{f_i} \quad (8.2.4b)$$

in the discrete case. Instead of a single likelihood equation one obtains a system of likelihood equations

$$\frac{\partial \log L}{\partial \theta_j} = 0 \quad (j = 1, \ldots, k), \quad (8.2.5)$$

which must be solved.

The multivariate case can be treated in a similar way.

Example 8.2.1

Suppose that the population distribution function is normal with unknown mean α and variance 1. The likelihood function is then given by

$$L(X_1, \ldots, X_n | \alpha) = (2\pi)^{-n/2} \exp\left[-\tfrac{1}{2} \sum_{j=1}^{n} (X_j - \alpha)^2\right],$$

and the likelihood equation becomes

$$\frac{\partial \log L}{\partial \alpha} \equiv \sum_{j=1}^{n} (X_j - \alpha) = 0.$$

This yields the maximum likelihood estimate $\hat{\alpha} = \overline{X}$. (Since $\frac{\partial^2 \log L}{\partial^2 \alpha} = -n < 0$, this is the unique maximum.)

Example 8.2.2

Suppose that the population distribution function is a Poisson distribution with probabilities $P(X = k) = \frac{e^{-\lambda}\lambda^k}{k!}$ ($k = 0, 1, \ldots$; λ unknown). Assume that a sample of size n was taken and that the value j was observed f_j times where $f_j \geq 0$, $\sum_j f_j = n$. The likelihood function is then

$$L(X_1, \ldots, X_n | \lambda) = e^{-\lambda n} \frac{\lambda^{\sum_j j f_j}}{\prod_{f_j > 0} (j!)^{f_j}}.$$

The likelihood equation is then

$$\frac{\partial \log L}{\partial \lambda} \equiv -n + \frac{\sum_j j f_j}{\lambda} = 0,$$

and has the unique root $\hat{\lambda} = \dfrac{\sum_j j f_j}{n} = \overline{X}$.

Example 8.2.3

As an example for the multiparameter case, we assume that the population distribution function is normal with mean α and variance σ^2. The likelihood function is then

$$L(X_1, \ldots, X_n | \alpha, \sigma) = (\sigma\sqrt{2\pi})^{-n} \exp\left[-\frac{1}{2\sigma^2} \sum_{j=1}^{n} (X_j - \alpha)^2\right],$$

8.2 POINT ESTIMATION

and the likelihood equations are

$$\frac{\partial \log L}{\partial \alpha} \equiv \frac{1}{\sigma^2} \sum_{j=1}^{n} (X_j - \alpha) = 0,$$

$$\frac{\partial \log L}{\partial \sigma} \equiv -\frac{n}{\sigma} + \frac{1}{\sigma^3} \sum_{j=1}^{n} (X_j - \alpha)^2 = 0.$$

The maximum likelihood estimates are the solutions of these equations and are

$$\hat{\alpha} = \overline{X}, \qquad \hat{\sigma}^2 = s^2.$$

The second method of estimation is the method of moments, due to Karl Pearson.

Let $X_1, \ldots, X_n$ be a sample taken from a population whose population distribution function $F(x|\theta_1, \ldots, \theta_k)$ depends on k parameters $\theta_1, \ldots, \theta_k$. Suppose that the first k moments $\alpha_1, \ldots, \alpha_k$ of F exist. These moments depend in general on the unknown parameters. One computes the sample moments $a_1, \ldots, a_k$ and sets each sample moment equal to the corresponding population moment. In this way one obtains a system of k equations.

$$a_j(X_1, \ldots, X_n) = \alpha_j(\theta_1, \ldots, \theta_k) \qquad (j = 1, \ldots, k). \tag{8.2.6}$$

If these equations can be solved for the $\theta_1, \ldots, \theta_k$ one obtains estimates $\tilde{\theta}_1, \ldots, \tilde{\theta}_k$ for the unknown parameters. The method of moments yields estimates that are usually less efficient than maximum likelihood estimates, but the calculation of estimates by the method of moments is often much simpler.

Example 8.2.4

The method of moments, applied to a normal population with mean α and variance σ^2 yields the estimates $\tilde{\alpha} = \overline{X}$, $\tilde{\sigma}^2 = s^2$.

Example 8.2.5

Suppose that the population distribution is a rectangular distribution over the interval $(\mu - \rho, \mu + \rho)$. Its frequency function is then

$$p(x) = \begin{cases} \dfrac{1}{2\rho} & \text{if } |x - \mu| < \rho, \\ 0 & \text{otherwise.} \end{cases}$$

It is easily seen that the first (population) moment is $\alpha_1 = \mu$, while the second moment $\alpha_2 = \mu^2 + \dfrac{\rho^2}{3}$. Let $a_1 = \dfrac{1}{n}\sum_{j=1}^{n} X_j$ and $a_2 = \dfrac{1}{n}\sum_{j=1}^{n} X_j^2$ be the first and second sample moments computed from a sample $X_1, \ldots, X_n$. The equations for the estimates are then

$$a_1 = \mu,$$

$$a_2 = \mu^2 + \dfrac{\rho^2}{3}.$$

These yield the estimates $\tilde{\mu} = a_1$, $\tilde{\rho}^2 = 3s^2$, where $s^2 = a_2 - a_1^2$ is the sample variance.

8.3 Interval Estimation

The point estimates that we discussed in the preceding section have a serious disadvantage, which is caused by the fact that they assign a definite numerical value to the unknown parameter. The probability that the estimate equals the unknown parameter will, in general, be zero. We illustrate this situation by an example.

Let $X_1, \ldots, X_n$ be a sample from a normal population with mean α and variance $\sigma^2 = 1$ and let ε be an arbitrary number. Then

$$P(\overline{X}_n = \alpha) \leq P\left(\alpha - \dfrac{\varepsilon}{\sqrt{n}} \leq \overline{X}_n \leq \alpha + \dfrac{\varepsilon}{\sqrt{n}}\right).$$

8.3 INTERVAL ESTIMATION

Since

$$P(-\varepsilon \leq (\overline{X}_n - \alpha)\sqrt{n} \leq \varepsilon) = \frac{1}{\sqrt{2\pi}} \int_{-\varepsilon}^{\varepsilon} \exp\left(-\frac{y^2}{2}\right) dy, \quad (8.3.1)$$

we see that

$$P(\overline{X}_n = \alpha) \leq \varepsilon \sqrt{\frac{2}{\pi}}.$$

Since ε can be chosen arbitrarily small we see that $P(\overline{X}_n = \alpha) = 0$.

It is therefore often preferable to use as an estimate a random interval and to select it in such a way that one can compute the probability that it covers the unknown parameter. We see from tables of the normal distribution (see Table IA of Appendix D) that

$$\frac{1}{\sqrt{2\pi}} \int_{-1.96}^{1.96} \exp\left(-\frac{y^2}{2}\right) dy = 0.95.$$

If we put in (8.3.1) $\varepsilon = 1.96$ we obtain

$$P(-1.96 \leq (\overline{X}_n - \alpha)\sqrt{n} \leq 1.96) = 0.95.$$

This can be rewritten as

$$P\left(\overline{X}_n - \frac{1.96}{\sqrt{n}} \leq \alpha \leq \overline{X}_n + \frac{1.96}{\sqrt{n}}\right) = 0.95. \quad (8.3.2)$$

We see therefore that the probability is 0.95 that the random interval $\left(\overline{X}_n - \frac{1.96}{\sqrt{n}}, \overline{X}_n + \frac{1.96}{\sqrt{n}}\right)$ contains (covers) the unknown parameter α. Clearly, this is also a method of estimation; it is called estimation by confidence interval.

Estimation by intervals requires the selection of two statistics, the first for the left, the second for the right endpoint of the interval. We can formally describe interval estimation in the following manner:

Let $\theta_1^*(X_1, \ldots, X_n)$ and $\theta_2^*(X_1, \ldots, X_n)$ be two statistics such that $\theta_1^* < \theta_2^*$ for all samples. The interval $[\theta_1^*, \theta_2^*]$ is said to be a confidence interval for the unknown parameter θ if the probability that it covers the parameter has a preassigned value $1 - \frac{p}{100}$.

That is, if for a given p $(0 < p < 100)$, the relation

$$P(\theta_1^* \leq \theta \leq \theta_2^*) = 1 - \frac{p}{100} \qquad (0 < p < 100) \qquad (8.3.3)$$

holds, we say that the interval $[\theta_1^*, \theta_2^*]$ is a $p\%$ confidence interval for the parameter θ. The positive number $1 - \frac{p}{100}$ is called its confidence coefficient.†

In the previous example, we gave a 5% confidence interval for the unknown mean α of a normal population with known variance $\sigma^2 = 1$. The confidence coefficient was 95%.

It is appropriate to make a few remarks concerning the meaning of interval estimation. First we consider the situation before a sample is taken. The endpoints θ_1^* and θ_2^* of the confidence interval are random variables, the interval $[\theta_1^*, \theta_2^*]$ is a random interval. The statement that the probability of its covering the unknown parameter is $1 - \varepsilon$ is therefore meaningful. However, after the sample is drawn, θ_1^* and θ_2^* are known numbers, the realization of the experiment could yield the values $\theta_1^* = 3$, $\theta_2^* = 5$ and the statement that $3 \leq \theta \leq 5$ contains no random variables and one can not talk in a meaningful way about its probability. Any statement about the probability that a confidence interval covers θ must be made before the sample is taken. The usefulness of the statement (8.3.3) is due to the law of large numbers. If we repeat the experiment a large number of times and assert before each measurement or observation that $[\theta_1^*, \theta_2^*]$ covers θ, then our statement will, in the long run, be wrong only in $p\%$ of the cases.

We give next a few examples for the determination of confidence intervals for the parameters of normal populations.

---

† The random variables θ_1^* and θ_2^* are called the confidence limits while $\frac{p}{100}$ is also called the confidence level.

8.3 INTERVAL ESTIMATION

We consider first a sample $X_1, \ldots, X_n$ taken from a normal population with mean α and variance σ^2. We know (Theorem 7.3.5) that the statistic

$$t = \frac{\sqrt{n-1}(\overline{X} - \alpha)}{s}$$

has Student's distribution with $n - 1$ degrees of freedom. Let t_p be the $p\%$ value of this distribution, then

$$P\left(-t_p < \sqrt{n-1}\,\frac{\overline{X} - \alpha}{s} \leq t_p\right) = 1 - \frac{p}{100},$$

or equivalently

$$P\left(\overline{X} - t_p \frac{s}{\sqrt{n-1}} \leq \alpha \leq \overline{X} + t_p \frac{s}{\sqrt{n-1}}\right) = 1 - \frac{p}{100}.$$

This indicates that

$$\left[\overline{X} - t_p \frac{s}{\sqrt{n-1}},\; \overline{X} + t_p \frac{s}{\sqrt{n-1}}\right]$$

is a confidence interval for the unknown mean α with confidence coefficient $(100 - p)\%$.

In the second example, we give a confidence interval for the standard deviation σ and assume again that the sample comes from a normal population with mean α and variance σ^2. Let $X_1, \ldots, X_n$ be a sample from this population and let s^2 be the variance of this sample. We know from Theorem 7.3.4 that $\frac{ns^2}{\sigma^2}$ has a chi-square distribution with $n - 1$ degrees of freedom. Select a positive number p ($0 < p < 100$) and let p_1 and p_2 be two positive numbers such that $p_1 < p_2$ and $p_1 + p_2 = p$. Let $\chi^2_{p_1}$ (respectively $\chi^2_{p_2}$) be the $p_1\%$ (respectively $p_2\%$) value of the chi-square distribution with $n - 1$ degrees of freedom. Then

$$P\left(\chi^2_{p_2} \leq \frac{ns^2}{\sigma^2} \leq \chi^2_{p_1}\right) = 1 - \frac{p}{100},$$

or equivalently

$$P\left(\frac{s\sqrt{n}}{\chi_{p_1}} \leq \sigma \leq \frac{s\sqrt{n}}{\chi_{p_2}}\right) = 1 - \frac{p}{100},$$

so that

$$\left[\frac{s\sqrt{n}}{\chi_{p_1}}, \frac{s\sqrt{n}}{\chi_{p_2}}\right]$$

is a confidence interval for the standard derivation with confidence coefficient $(100 - p)\%$.

Here we are still free to choose p_1 or p_2; therefore, it is clear that we can construct infinitely many confidence intervals for σ with a confidence coefficient of $(100 - p)\%$. Thus, it is necessary to make a choice among these intervals. As a general rule, we can state that it is desirable to have as short a confidence interval as possible. The selection of a short—or possibly the shortest—confidence interval is an interesting problem that we cannot treat here.

Next we assume that two independent samples $X_1, \ldots, X_n$ and $Y_1, \ldots, Y_m$ can be observed. We assume that they come from normal populations having the same variance σ^2, and we wish to investigate whether the population means are also equal. This can be done by constructing a confidence interval for the two population means. Let α_1 (respectively α_2) be the mean of the population from which $X_1, \ldots, X_n$ (respectively $Y_1, \ldots, Y_m$) was drawn. Let $\bar{X}$ and s_1^2 (respectively $\bar{Y}$ and s_2^2) be the sample mean and the sample variance of the sample $X_1, \ldots, X_n$ (respectively $Y_1, \ldots, Y_m$). It can be shown† that $\dfrac{ns_1^2 + ms_2^2}{\sigma^2}$ has a chi-square distribution with $(n + m - 2)$ degrees of freedom while $\bar{X} - \bar{Y} - (\alpha_1 - \alpha_2)$ has a normal distribution with mean zero and variance $\sigma^2\left(\dfrac{1}{n} + \dfrac{1}{m}\right)$. We introduce the statistic

$$t = \sqrt{\frac{(n + m - 2)nm}{n + m}} \frac{\bar{X} - \bar{Y} - (\alpha_1 - \alpha_2)}{\sqrt{ns_1^2 + ms_2^2}},$$

---

† The proof of this statement requires the same prerequisites as the proof of Theorems 7.3.3 and 7.3.4, so it is impossible to give it here. See [*1*, p. 388] or [2, p. 188].

8.3 INTERVAL ESTIMATION

and see from Theorem 6.4.1 that t has Student's distribution with $n + m - 2$ degrees of freedom. It follows then that

$$\left[\overline{X} - \overline{Y} - t_p \sqrt{\frac{(n + m)(ns_1^2 + ms_2^2)}{nm(n + m - 2)}},\right.$$

$$\left.\overline{X} - \overline{Y} + t_p \sqrt{\frac{(n + m)(ns_1^2 + ms_2^2)}{nm(n + m - 2)}}\right]$$

is a confidence interval for the difference $\alpha_1 - \alpha_2$ of the unknown means. The confidence coefficient is $(100 - p)\%$.

Example 8.3.1 †

Two methods were used in the study of the latent heat of fusion of ice. Both Method A (an electrical method) and Method B (a method of

Table 8.1

Change in Total Heat from $-0.72°C$ to $0°C$[a]

Method A	Method B
79.98	80.02
80.04	79.94
80.02	79.98
80.04	79.97
80.03	79.97
80.03	80.03
80.04	79.95
79.97	79.97
80.05	
80.03	
80.02	
80.00	
80.02	

[a] Values are in calories per gram mass.

† This example is taken from [3, p. 3.23].

mixtures) were conducted with specimens cooled to $-0.72°C$. The data represent the change in total heat from $-0.72°C$ to water at $0°C$ in calories per gram mass (see Table 8.1).

Determine a 95% confidence interval for the difference of the two means.

Let $\overline{X}$ be the mean determined by Method A and $\overline{Y}$ the mean determined by Method B. One sees that $\overline{X} = 80.02$, $\overline{Y} = 79.98$, $n = 13$, $m = 8$. At 19 degrees of freedom the 5% value of Student's distribution is 2.09 and we obtain the 95% confidence interval (0.015, 0.065).

Example 8.3.2

Ten units of rocket powder, selected at random from a lot, were tested and their burning times (in seconds) were recorded.

50.7	69.8
54.9	53.4
54.3	66.1
44.8	48.1
42.2	34.5

The standard deviation of these measurements is used as a measure of their variability. Find a two-sided 90% confidence interval for the standard deviation.

A simple computation yields $\overline{X} = 52.0$, $s^2 = 1078.0$, $s = 32.83$. We see from the table of the chi-square distribution that the 95% value (at 9 degrees of freedom) is 3.325, and that the 5% value is 16.919. Therefore, $\chi_{0.95} = \sqrt{16.919} = 4.01$, and that $\chi_{0.05} = \sqrt{3.325} = 1.82$. This yields the 90% confidence interval [24.6, 54.1] for the standard deviation.

The construction of the confidence intervals that we presented used the knowledge of the distribution of certain statistics. In the absence of this information one can still construct confidence intervals based on large samples. This is done by using the normal approximation, which is asymptotically valid in many cases.

8.4 PROBLEMS

We mention as a simple example the estimation of the mean α of a population whose distribution function is not normal, but whose standard deviation σ is supposed to be known.

Tables for the $p\%$ values of the standardized normal distribution are available (see Table IA of Appendix D), these give the value λ_p as a function of p, where λ_p is determined by the relation

$$1 - 2[1 - \Phi(\lambda_p)] = \frac{1}{\sqrt{2\pi}} \int_{-\lambda_p}^{\lambda_p} \exp\left(-\frac{y^2}{2}\right) dy = 1 - \frac{p}{100}.$$

The statistic $\sqrt{n}\,\dfrac{\overline{X} - \alpha}{\sigma}$ has asymptotically a standardized normal distribution so that

$$P\left(-\lambda_p \leq \sqrt{n}\,\frac{\overline{X} - \alpha}{\sigma} \leq \lambda_p\right) = 1 - \frac{p}{100}.$$

This equation indicates that

$$\left[\overline{X} - \lambda_p \frac{\sigma}{\sqrt{n}},\ \overline{X} + \lambda_p \frac{\sigma}{\sqrt{n}}\right]$$

is asymptotically a confidence interval for the unknown mean α with confidence coefficient equal to $1 - \dfrac{p}{100}$.

If the standard deviation σ is not known then we could approximate σ by s and obtain as an approximation the confidence interval

$$\left[\overline{X} - \lambda_p \frac{s}{\sqrt{n}},\ \overline{X} + \lambda_p \frac{s}{\sqrt{n}}\right].$$

8.4 Problems

1. Let $X_1, \ldots, X_n$ be a sample from a population and suppose that the common distribution of the X_j has finite variance. Let a_{nk} ($k = 1, \ldots, n$) be real numbers such that $a_{nk} \geq 0$ and $\sum_{k=1}^{n} a_{nk} = 1$.

Consider the weighted means $\theta_n = \sum_{k=1}^{n} a_{nk} X_k$ and show that

(a) θ_n is an unbiased estimate of the population mean.

(b) The θ_n are a consistent sequence of estimates for the population mean provided that $\lim_{n \to \infty} \sum_{k=1}^{n} a_{nk}^2 = 0$.

2. Let $X_1, \ldots, X_n$ be a sample from a population whose distribution function has finite variance. Let $\{a_j\}$ be a sequence of real numbers which has a positive lower and a finite upper bound. Show that the sequence

$$\theta_n^* = \frac{a_1 X_1 + \cdots + a_n X_n}{a_1 + \cdots + a_n}$$

is a consistent sequence of estimates for the population mean.

3. Let $X_1, \ldots, X_n$ be a sample from a Poisson population and show that $\overline{X}_n$ and $\dfrac{n s_n^2}{n-1}$ are both sequences of unbiased estimates for the population parameter λ.

4. Let $X_1, \ldots, X_n$ be a sample from a population and suppose that the common frequency function of the $X_1, \ldots, X_n$ belongs to a gamma distribution with parameters θ and λ. Estimate these parameters by the method of moments.

5. Let $X_1, \ldots, X_n$ be a sample from a population and assume that the common frequency function of the $X_1, \ldots, X_n$ belongs to a gamma distribution with known parameter λ but unknown parameter θ. Find the maximum likelihood estimate for θ.

6. Suppose that the common frequency function of the random variables of a sample $X_1, \ldots, X_n$ is given by

$$p(x) = \begin{cases} \theta e^{-\theta x} & \text{for } x > 0, \\ 0 & \text{for } x < 0. \end{cases}$$

Estimate the parameter θ by the maximum likelihood method.

8.4 PROBLEMS

7. Suppose that $X_1, \ldots, X_n$ is a sample from a population whose frequency function is given by
$$p(x) = \begin{cases} \theta e^{-\theta x} & \text{for } x > 0, \\ 0 & \text{for } x < 0. \end{cases}$$
Estimate the parameter θ by the method of moments.

8. Let $X_1, \ldots, X_n$ be a sample from a population whose frequency function is given by
$$p(x) = \begin{cases} \theta e^{-\theta(x-\mu)} & \text{for } x > \mu, \\ 0 & \text{for } x < \mu. \end{cases}$$
Estimate the parameters θ and μ by the method of moments.

9. Let $X_1, \ldots, X_n$ be a sample from a population whose frequency function is given by
$$p(x) = \begin{cases} \dfrac{|x|}{r^2} & \text{if } |x| < r, \\ 0 & \text{if } |x| > r. \end{cases}$$
Estimate r by the method of moments.

10. Let $p(x)$ be the frequency function of Simpson's (triangular) distribution, that is
$$p(x) = \begin{cases} \dfrac{k+x}{k^2} & \text{if } -k \leq x \leq 0, \\ \dfrac{k-x}{k^2} & \text{if } 0 \leq x \leq k, \\ 0 & \text{if } |x| \geq k. \end{cases}$$
Suppose that a sample of size n is taken from a population that has this frequency function and
   (a) estimate the parameter k by the method of moments
   (b) determine—without solving it—the likelihood equation.

11. The breaking strengths (in pounds) of five pieces of rope are 540, 485, 525, 590, 560. Use the maximum likelihood method to estimate the breaking strength, assuming a normal distribution.

12. Ten washers are taken at random from a large group and their thickness is measured in inches. The following values are obtained:

0.123	0.132
0.124	0.123
0.126	0.123
0.129	0.129
0.120	0.128

   (a) Estimate the average thickness of the washers.

   (b) Find a 95% confidence interval for their mean thickness.

13. Determine a 99% confidence interval for Example 8.3.1.

14. Use the data of Problem 9 of Chapter 7 and determine a 90% confidence interval for the difference of the means determined by Methods A and B.

15.† Two investigators (A and B) obtained specimen cores of the concrete in a poured slab in order to determine its compressive strength. The results (in pounds per square inch) are given in Table 8.2.

Table 8.2

Compressive Strength of Concrete Samples[a]

A	B
3128	1939
3219	1697
3244	3030
3037	2424
	2020
	2909
	1815
	2020
	2310

[a] Values given in pounds per square inch.

---

† This problem is taken from [3, p. 3.26].

Determine a 95% confidence interval for the difference of the means obtained by the two investigators.

16.† The capacities (in ampere hours) of 10 batteries were recorded

146	143
141	138
135	137
142	142
140	136

Find a two-sided 10% confidence interval for the standard deviation.

17. Let $\{\theta_n^*\}$ be a sequence of biased estimates for the parameter θ and denote the bias of θ_n^* by b_n. Suppose that $\lim_{n \to \infty} \text{Var}(\theta_n^*) = 0$ and also $\lim_{n \to \infty} b_n = 0$. Show that the $\{\theta_n^*\}$ form a sequence of consistent estimates.

References

1. H. Cramér, "Mathematical Methods of Statistics." Princeton Univ. Press, Princeton, New Jersey, 1946.
2. H. Cramér, "The Elements of Probability Theory and Some of Its Applications." Wiley, New York, 1955.
3. M. Natrella, "Experimental Statistics (NBS Handbook 99)." U.S. Govt. Printing Office, Washington, D.C., 1963.

---

† This problem is taken from [3, p. 4.1].

9 | Testing Hypotheses

The problems discussed in this chapter are motivated by questions that arise in a great variety of fields, for example, in industrial, agricultural, medical, or biological research. In all these—and in many other areas—the subject matter specialist often must make a decision in the face of uncertainty. This decision consists in choosing one of two possible actions, say A or B. For this decision, a number of data, obtained through observation or experimentation, are available. These data must be used in a rational way as a basis for the decision. In such a situation the construction of a probabilistic model is appropriate and helpful.

In this chapter we discuss first in an abstract way the underlying statistical principles and develop also some useful techniques.

9 TESTING HYPOTHESES

9.1 Statistical Hypotheses

Let us assume that a probabilistic model is available for the phenomenon to be investigated. A hypothesis can be formulated in the framework of the model and the available data must be used either to accept or reject the hypothesis.

The terms "accept" or "reject" a hypothesis are universally employed. One should, however, discard any deeper meaning that the common usage of these terms might suggest. To accept a hypothesis does not mean that one has established that it is true. It only means that one has decided to take a certain action, say A. Similarly, in rejecting a hypothesis one does not claim that it is false, one has only decided to take a different action, say B. In either case, there exists the possibility that a wrong decision was made. It is the aim of statistical theory to control the probabilities of erroneous decisions.

The model usually has the following structure: A certain quantity is assumed to be a random variable with an unknown distribution function. It is, however, assumed that the functional form of the distribution function is known, but that it depends on certain unknown parameters $\theta_1, \ldots, \theta_k$. We denote then the distribution function by $F(x \mid \theta_1, \ldots, \theta_k)$ and the set of possible values of the parameter by θ. We can consider θ as a k-dimensional space or a subset of a k-dimensional space. In either case we call θ the parameter space. The hypothesis is then formulated as an assumption concerning the unknown parameter (or parameters) and is tested on the basis of the information contained in a sample $X_1, \ldots, X_n$ obtained by observation or by experimentation.

There are two types of statistical hypotheses: The hypothesis can specify the parameters completely; it is then called a simple hypothesis. Alternatively it is possible that a hypothesis does not determine uniquely the parameters; in this case it is said to be a composite hypothesis.

Example 9.1.1

We assume that we wish to study the breaking strength X of a certain type of cable by carrying out 100 experiments. We assume that

9.1 STATISTICAL HYPOTHESES

X is normally distributed about an average breaking strength α with a standard deviation σ. Then

$$F(x) = \frac{1}{\sigma\sqrt{2\pi}} \int_{-\infty}^{x} \exp\left[-\frac{1}{2\sigma^2}(y-\alpha)^2\right] dy,$$

while the sample size $n = 100$. An assumption that the breaking strength is 30 psi with a standard deviation of 3 psi, is a simple hypothesis. The assumption that the breaking strength is greater than 30 psi, while the standard deviation is 3 psi is a composite hypothesis.

Let us consider in this example the simple hypothesis that the breaking strength is 30 psi, with a standard deviation of 3 psi. If the available evidence suggests that this hypothesis is reasonable, then we shall buy the cable (action A); if the evidence indicates that the assumption of the hypothesis is unreasonable, then we shall not buy the cable (action B).

We mention briefly two other examples: An engineer may have to decide which of two manufacturing processes should be used in producing an item. An agricultural scientist may have to decide which of two fertilizers should be used.

Let us suppose that H_0 is the hypothesis to be tested.† One has to decide on the basis of a sample $X_1, \ldots, X_n$ whether or not the null hypothesis should be rejected. It is convenient to formulate the problem by representing the n observations $X_1, \ldots, X_n$ as a point of an n-dimensional Euclidean space. This space is called the sample space. To solve the problem of testing we select a region R in the sample space such that the probability of the sample point $\mathbf{X} = (X_1, \ldots, X_n)$ falling into R is small in case the null hypothesis is true. We agree to reject H_0 if the sample point falls into R. The region R is called the critical region (or the region of rejection). We write $P(\mathbf{X} \in R \mid H_0)$, or, for the sake of brevity, $P(R \mid H_0)$, for the probability that $\mathbf{X} \in R$ if H_0 is true, and select R so that $P(R \mid H_0) = \alpha$. The constant α is called the size of the critical region or the significance level of the test. The first step is the selection of a (suitably small) significance level α, then a critical region R of size

---

† It is customary to call the hypothesis to be tested the null hypothesis.

α has to be chosen. The sample is then taken, and the decision is made to reject or to accept H_0 in the manner just described. This procedure is not foolproof; it does not guarantee that the conclusion will always be correct. The acceptance of the null hypothesis does not carry the force of a mathematical proof; we can assure only that the conclusion will be correct with a certain probability. This means that we can expect to arrive at a correct conclusion in a precisely stated percentage of the cases if we use the same test a large number of times.

Let us consider a simple example. Suppose that $F(x|\theta)$ is a normal distribution with unknown mean θ but known variance $\sigma^2 = 1$. The null hypothesis H_0 to be tested is the statement that the mean is zero, $\theta = 0$. It is reasonable to reject H_0 if the observed sample mean is too large in absolute value. The sample mean $\bar{X} = \frac{1}{n}(X_1 + \cdots + X_n)$ has, under the null hypothesis,† a normal distribution with mean 0 and variance $\frac{1}{n}$. We see from the tables of the standardized normal distribution (Table I of Appendix D) that $P\left(|\bar{X}| > \frac{1.960}{\sqrt{n}}\right) = 0.05$. We can therefore use the set $\left[|\bar{X}| > \frac{1.960}{\sqrt{n}}\right]$ as a critical region of size 5%. However, this region is not the only critical region at the 5% level of significance. It is easily seen that the regions $[\bar{X} > 1.751/\sqrt{n}$ or $\bar{X} < -2.327/\sqrt{n}]$ or $[\bar{X} > 1.645/\sqrt{n}]$ or $[\bar{X} < -1.645/\sqrt{n}]$ are also 5% critical regions. Of course it is possible to construct infinitely many critical regions of a given size. It is therefore necessary to develop principles for the selection of an optimal region of a given size.

9.2 The Power of a Test

In order to arrive at a rational method for the selection of a critical region, it is necessary to consider the possible errors that one can

---

† The phrase, "under the null hypothesis" means under the assumption that the null hypothesis is true.

9.2 THE POWER OF A TEST

commit. One can reject the null hypothesis H_0 even if the statement of the hypothesis is true—such an error is called a Type-I error. Or one can fail to reject H_0 when it is false (that is, accept it erroneously)—such an error is called a Type-II error. Table 9.1 surveys these types of errors:

Table 9.1

Type-I and Type-II Errors

True situation	Statement made	
	Accept H_0	Reject H_0
H_0 true	Correct	Type-I error
H_0 false	Type-II error	Correct

Let H_0 be the simple hypothesis that $\theta = \theta_0$ and let R be a critical region of size α. We write $P(R|\theta)$ for the probability that the sample point is located in R, if the true value of the unknown parameter is θ. Then

$$P(R|\theta_0) = \alpha \qquad (9.2.1)$$

is the probability of committing a Type-I error. The size α of the critical region controls the Type-I error. In general, one has infinitely many critical regions of a given size α. It is reasonable to select from these regions one that makes the probability of a Type-II error small. Let H_1 be an alternative hypothesis† to H_0. Then

$$P(\mathbf{X} \in R | H_1)$$

is the probability that H_0 is rejected if the alternative hypothesis H_1 is true. Then

$$1 - P(\mathbf{X} \in R | H_1)$$

---

† For instance $\theta \neq \theta_0$. An alternative could also be a less general statement such as $\theta > \theta_0$ or could consist in the specification of a value for θ that is different from θ_0, say $\theta = \theta_1 \neq \theta_0$.

is the probability of failing to reject H_0 when it is false. This means that $1 - P(\mathbf{X} \in R | H_1)$ is the probability of committing a Type-II error; this probability depends on the alternative hypothesis. We write

$$P(\mathbf{X} \in R | H_1) = P(R | H_1) = \pi(H_1) \qquad (9.2.2)$$

and call $\pi(H_1)$ the power of the test with respect to the alternative H_1. The power of a test with respect to an alternative hypothesis is the probability of not committing a Type-II error if the alternative is true. We write $\pi_R(H_1)$ if we wish to put into evidence the critical region used in constructing the test.

Let us consider again a normal population with unknown mean θ and known standard deviation σ. Let $\theta = \theta_0$ be the null hypothesis H_0, we select as the critical region the region in the sample space where

$$\frac{\sqrt{n}|\bar{X} - \theta_0|}{\sigma} \geq \lambda_p. \qquad (9.2.3)$$

where λ_p is the $p\%$ value of the standardized normal distribution. This is a critical region of size $\frac{p}{100}$. The hypothesis H_0 is accepted if

$$\frac{\sqrt{n}|\bar{X} - \theta_0|}{\sigma} < \lambda_p$$

and is rejected if $\frac{\sqrt{n}|\bar{X} - \theta_0|}{\sigma} \geq \lambda_p$. As an alternative hypothesis, we consider the assumption that the population mean is equal to θ where $\theta \neq \theta_0$. The power of the test against this alternative is then

$$\pi(\theta) = P(R | \theta),$$

that is, the probability that the sample point falls into R if θ is the mean of the population. This is obviously a function of θ, which is called the power function $\pi(\theta)$ of the test. We determine next $\pi(\theta)$. Let

$$Y = \frac{\sqrt{n}(\bar{X} - \theta)}{\sigma}, \qquad (9.2.4a)$$

9.2 THE POWER OF A TEST

and
$$d = \frac{\sqrt{n}\,(\theta - \theta_0)}{\sigma}, \tag{9.2.4b}$$

then
$$\frac{\sqrt{n}}{\sigma}(\bar{X} - \theta_0) = Y + d. \tag{9.2.5}$$

If the alternative hypothesis is true, then Y is normally distributed with zero mean and unit variance, while d is a constant. We have, at the $p\%$ level,

$$P(R|\theta) = P(\bar{X} \in R|\theta) = P\left(\left|\frac{(\bar{X} - \theta_0)\sqrt{n}}{\sigma}\right| \geq \lambda_p \,\bigg|\, \theta\right).$$

Using (9.2.5) we see that

$$\pi(\theta) = P(R|\theta) = P(Y \geq \lambda_p - d|\theta) + P(Y \leq -\lambda_p - d|\theta).$$

Under the alternative hypothesis Y has a standardized normal distribution and we see that

$$\pi(\theta) = 1 - \Phi(\lambda_p - d) + \Phi(-\lambda_p - d).$$

We substitute for d from (9.2.4b) and obtain

$$\pi(\theta) = 1 - \Phi\left(\lambda_p - \frac{\sqrt{n}(\theta - \theta_0)}{\sigma}\right) + \Phi\left(-\lambda_p - \frac{\sqrt{n}(\theta - \theta_0)}{\sigma}\right). \tag{9.2.6}$$

As usual,
$$\Phi(x) = \frac{1}{\sqrt{2\pi}} \int_{-\infty}^{x} \exp\left(-\frac{y^2}{2}\right) dy.$$

The following relations are an immediate consequence of (9.2.6)

$$\pi(\theta_0) = \alpha \tag{9.2.7a}$$

$$\lim_{\theta \to \infty} \pi(\theta) = 1 \tag{9.2.7b}$$

$$\lim_{\theta \to -\infty} \pi(\theta) = 1, \qquad (9.2.7c)$$

$$\lim_{n \to \infty} \pi(\theta) = 1. \qquad (9.2.7d)$$

We note that $\pi(\theta)$ is a continuous function of θ, therefore

$$\lim_{\theta \to \theta_0} \pi(\theta) = \alpha,$$

this means that our test does not discriminate well against alternatives that are close to the null hypothesis. On the other hand, we see from (9.2.7d) that for a given alternative hypothesis θ the power of the test can be made as large as we wish by selecting a sufficiently large sample size.

Example 9.2.1

As a numerical example we compute $\pi(\theta)$ and draw its graph for the case $\theta_0 = 0$, $\sigma = 1$, $p = 0.05$, and $n = 4$.

θ	$\pi(\theta) = \pi(-\theta)$
0	0.05
0.05	0.051
0.10	0.055
0.15	0.060
0.20	0.068
0.25	0.079
0.30	0.092
0.40	0.126
0.50	0.170
1.00	0.516
1.50	0.851
2.00	0.979
2.50	0.999

We now return to the consideration of tests for a simple hypothesis. Let R_1 and R_2 be two critical regions of size α for testing the null hypothesis

$$H_0 : \theta = \theta_0$$

9.2 THE POWER OF A TEST

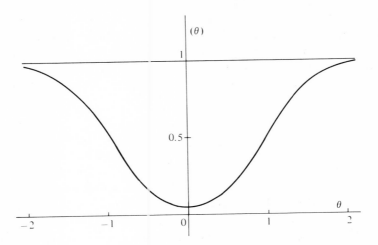

Figure 9.1. Power curve.

against the alternative

$$H_1 : \theta = \theta_1.$$

Suppose that the powers $\pi_{R_1}(\theta_1)$ and $\pi_{R_2}(\theta_1)$ satisfy the inequality

$$\pi_{R_1}(\theta_1) > \pi_{R_2}(\theta_1).$$

In such a case, we shall prefer the region R_1 to the region R_2. If the power curves $\pi_{R_1}(\theta)$ and $\pi_{R_2}(\theta)$ of the regions R_1 and R_2, respectively, satisfy the relation

$$\pi_{R_1}(\theta) \geq \pi_{R_2}(\theta)$$

for all $\theta \in \theta$ with the sharp inequality valid for at least some θ then we say that R_1 is uniformly more powerful than R_2. In such a situation one will select the region R_1 for the test. It is possible to carry this consideration one step further. Let $\mathfrak{R}_\alpha$ be the set of all critical regions of size α for testing the null hypothesis $H_0 : \theta = \theta_0$. If there exists a critical region $R^* \in \mathfrak{R}_\alpha$ such that

$$\pi_{R^*}(\theta) \geq \pi_R(\theta)$$

for all $R \in \mathfrak{R}_\alpha$ and all $\theta \in \boldsymbol{\theta}$ then R^* is said to be a uniformly most powerful region for testing H_0. Unfortunately uniformly most powerful critical regions do not always exist. Interesting results concerning the existence of uniformly most powerful critical regions were obtained by Neyman and Pearson. The study of these results would, however, exceed the scope of this book.

9.3 The t-Test

In this section we give an example for testing a composite hypothesis. This example will also be used to show the close connection between testing hypotheses and estimation by confidence intervals.

We consider a normal population with mean θ and variance σ^2 where both, θ and σ^2, are unknown. The problem is to test the hypothesis $\theta = \theta_0$ at the significance level α against the alternative $\theta \neq \theta_0$ on the basis of a sample $X_1, \ldots, X_n$ of size n. The null hypothesis is obviously a composite hypothesis. The set of alternatives is $\boldsymbol{\theta} = \{(\theta, \sigma^2): -\infty < \theta < +\infty, \theta \neq \theta_0, \sigma^2 > 0\}$. If the null hypothesis is true, the statistic†

$$t = \frac{\sqrt{n-1}(\overline{X} - \theta_0)}{s}$$

has Student's distribution with $n - 1$ degrees of freedom. It seems to be reasonable to reject the null hypothesis if the value of $|t|$ is too large. This leads to the critical region

$$\left| \sqrt{n-1} \frac{(\overline{X} - \theta_0)}{s} \right| > t_p, \qquad (9.3.1)$$

where t_p is the $p\%$ value of Student's distribution. The size of the critical region (9.3.1) is $\alpha = \dfrac{p}{100}$.

---

† See Section 8.3.

9.3 THE t-TEST

The critical region (9.3.1) is the complement of the confidence interval, which we derived on page 165. The confidence coefficient of this interval is $(100 - p)\%$. Thus the problems of testing hypotheses concerning a parameter and the construction of confidence intervals are very closely related.

Example 9.3.1

A manufacturer produces a certain type of cable and claims that it has a breaking strength of 35 psi. Ten experiments were performed to test this claim; the mean breaking strength observed was 30.5 psi. with a standard deviation of 3.3 psi. Is the claim of the manufacturer justified? Assuming that the breaking strength is normally distributed about the true mean we use the t-test for the null hypothesis that the mean $\alpha = 35$. We have

$$t = \sqrt{n-1}\,\frac{\overline{X} - \alpha}{s} = \sqrt{9}\,\frac{30.5 - 35}{3.3} = -4.09.$$

The 5% value of Student's t is 2.262 so that the observed value is too large. The null hypothesis is therefore rejected and the manufacturer's claim appears to be unjustified.

It is also possible to use the t-test in a number of other situations, for instance to test the equality of two means of two normal populations having the same variance.

Example 9.3.2

Use the data of Example 8.3.1 to test whether the means obtained by Methods A and B, respectively, are equal.

We saw in Section 8.3 (page 166) that the quantity

$$t = \sqrt{\frac{(n + m - 2)nm}{(n + m)(ns_1^2 + ms_2^2)}}\,[\overline{X} - \overline{Y} - (\alpha_1 - \alpha_2)]$$

has a Student's distribution with $n + m - 2$ degrees of freedom. The null hypothesis is $\alpha_1 = \alpha_2$, under the null hypothesis we obtain from the data $t = 3.30$. The 5 % (1 %) value of Student's distribution at 19 degrees of freedom is 2.09 (2.861) so that we must reject the hypothesis that the means are equal.

9.4 Nonparametric Methods

The statistical methods that we have discussed so far were all based on the assumption that the functional form of the population distribution function is known, but that it contains some unknown parameters. One either had to estimate these parameters or had to test some hypothesis concerning them.

However, the statistician also encounters situations where he feels that it would not be justified to assume that the functional form of the population distribution function is known. Methods have been developed which do not assume that the form of the population distribution function is known. Most of these need only the assumption that the population has an absolutely continuous distribution function. These methods are called nonparametric (or distribution free) methods. They use less *a priori* information about the population than the parametric methods, therefore, it is not surprising that they are less efficient. In this section, we discuss only a few examples; it is not our aim to treat this subject in any detail or to mention many of the existing techniques.

We consider the problem of deciding whether two frequency functions are identical. This decision should be made on the basis of two samples of equal size n. Under certain additional restrictions, namely if one assumes that both populations are normal and have a common variance, Student's t-test can be used for this purpose. The first test to be discussed in this section is called the sign test; its importance lies in the fact that the conditions of Student's test need not be satisfied. One assumes only that n independent pairs of observations were made and that the two observations of each pair were obtained under similar

9.4 NONPARAMETRIC METHODS

conditions. However, one can admit the possibility† that different pairs were observed under different conditions.

Let $(X_1, Y_1), (X_2, Y_2), \ldots, (X_n, Y_n)$ be the n pairs of observations and suppose that $p_{1,j}(x)$ is the frequency function of X_j, and $p_{2,j}(x)$ is the frequency function of Y_j. The null hypothesis is

$$H_0 : p_{1,j}(x) \equiv p_{2,j}(x) \qquad (j = 1, 2, \ldots, n). \tag{9.4.1}$$

We consider the n differences $X_j - Y_j$ $(j = 1, \ldots, n)$ between the observations of each pair. If the null hypothesis is correct the two sample values will be different,‡ moreover

$$P(X_j - Y_j > 0) = P(X_j - Y_j < 0) = \tfrac{1}{2}. \tag{9.4.2}$$

We introduce the random variables

$$Z_j = \begin{cases} 1 & \text{if } X_j - Y_j > 0, \\ 0 & \text{if } X_j - Y_j < 0. \end{cases}$$

The random variables Z_j have only two possible values and

$$P(Z_j = 1) = P(Z_j = 0) = \tfrac{1}{2} \qquad (j = 1, \ldots, n).$$

The random variables $Z_1, \ldots, Z_n$ are independent, and their sum $S_n = \sum_{j=1}^{n} Z_j$ is the number of positive differences $X_j - Y_j$. Under the null hypothesis S_n has a binomial distribution with probability $p = \tfrac{1}{2}$, and one can construct a test using tables of the binomial distribution. We determine an integer k such that

$$\sum_{j=k}^{n-k} \binom{n}{j} \left(\frac{1}{2}\right)^n \geq 1 - \alpha. \tag{9.4.3}$$

---

† This possibility is one of the valuable features of the sign test. The distributions may change from one pair of observations to the next. In such a situation the t-test cannot be used while the sign test is still applicable. Another good feature of the sign test is that it requires almost no computations.

‡ This follows from the absolute continuity of the distribution functions.

The null hypothesis is then rejected if either $S_n < k$ or $S_n > n - k$. The test with the critical region $[S_n < k$ or $S_n > n - k]$ is a two-sided test at the significance level α. It is also possible to construct one-sided tests. The critical region $[S_n < k]$ rejects H_0 if too few positive differences are observed, it has the significance level $\frac{\alpha}{2}$. Similarly, one can construct a test that rejects H_0 if two many positive differences are observed.

The condition that the distribution functions of the X_j and Y_j be absolutely continuous can be removed by a slight modification of the procedure. If the possibility of ties (that is, of $X_j = Y_j$ for some j's) is admitted, then it is best to disregard the ties and to proceed with the remaining observations as outlined above. It can be shown that this will result in a test with significance level not exceeding α.

Tables of the percentage points of the sign test are available (see Table IV of Appendix D).

Next we consider the oldest distribution-free test which was introduced by K. Pearson. The test deals with the following problem. A sample $X_1, \ldots, X_n$ from a population is given. One wishes to test the hypothesis that the population distribution function is a completely specified distribution function $F(x)$. To do this one divides the range of $F(x)$ into a certain number, say s, of nonoverlapping intervals or groups. Since the distribution function $F(x)$ is completely specified, it is possible to compute the probabilities p_j that an observation falls into the jth group. This probability must then be compared with the observed frequencies O_j of observations falling into this group. In order to arrive at a judgment whether the hypothesis that $F(x)$ is the distribution function that fits the data, one must introduce a suitable measure for the deviation between the theoretical distribution and the observed frequencies. Let $p_1, \ldots, p_s$ be the theoretical probabilities, and denote the observed frequencies in the s groups by $O_1, \ldots, O_s$. Then

$$\sum_{j=1}^{s} p_j = 1, \qquad \sum_{j=1}^{s} O_j = n. \qquad (9.4.4)$$

The expected number of observations in the jth group is, under the null

9.4 NONPARAMETRIC METHODS

hypothesis, equal to np_j. Therefore, one must construct a measure which compares np_j and O_j. Pearson proposed the statistic

$$\chi^2 = \sum_{j=1}^{s} \frac{(O_j - np_j)^2}{np_j} = \sum_{j=1}^{s} \frac{O_j^2}{np_j} - n. \qquad (9.4.5)$$

Let

$$Z_j = \frac{O_j - np_j}{\sqrt{np_j}} \quad (j = 1, \ldots, s),$$

then

$$\chi^2 = \sum_{j=1}^{s} Z_j^2,$$

where

$$\sum_{j=1}^{s} Z_j \sqrt{p_j} = 0.$$

The last equation follows from (9.4.4).

The random variables $O_1, \ldots, O_s$ have a multinomial distribution. Using this fact, one can obtain the limiting distribution of $Z_1, \ldots, Z_s$ as n tends to infinity. From this limiting distribution, it is possible to derive the asymptotic distribution of χ^2. One can show that the distribution of χ^2, as given by (9.4.5) tends to a chi-square distribution with $s - 1$ degrees of freedom as n tends to infinity. The necessary computations are tedious and require analytical tools that are beyond the scope of this book. Therefore, we must omit the proof of these results. A proof can be found in [1, pp. 416 ff].

We finally remark that a similar chi-square test of goodness of fit can be developed for the case where the distribution function $F(x)$ contains some unknown parameters that must be estimated from the sample. Chi-square tests of goodness of fit are also very useful in cases where a group of individuals is classified according to two characteristics and where one investigates whether these characteristics are independent. Problems of this kind occur frequently in genetics.

Example 9.4.1†

Ten pairs of measurements of reverse bias collector currents of two types of transistors were made. The results are given in Table 9.2.

Table 9.2

Reverse Bias Collector Currents

Transistor A	Transistor B
0.19	0.21
0.22	0.27
0.18	0.15
0.17	0.18
1.20	0.40
0.14	0.08
0.09	0.14
0.13	0.28
0.26	0.30
0.66	0.68

The observations of each pair were obtained under similar conditions, but the conditions may have varied from pair to pair. Do the data indicate a significant difference between the two types of transistors?

We apply the sign test. We have $n = 10$ observations (without ties) and $r = 3$ positive signs of the difference $A - B$. For $n = 10$, Table IV gives 0 at the 1% level, 1 at the 5 and 10% levels, and 2 at the 25% level. We conclude therefore that there is no significant difference between the two types of transistors.

Example 9.4.2

Clinical thermometers are classified into the following four categories: (1) nondefective, (2) defective—Class A (defects in glass, mark-

---

† Examples 9.4.1 and 9.4.2 are taken from [2, pp. 16.8 and 9.2] respectively.

ings, dimensional nonconformance), (3) defective—Class B (defects in mercury column), (4) defective—Class C (nonconformance to precision and accuracy requirements).

Over a period of time it has been found that thermometers produced by a certain manufacturer are distributed in the following average proportions:

Nondefective	87%
Class A	9%
Class B	3%
Class C	1%

A new lot of 1336 thermometers was submitted for inspection, and the result was the following classification:

Nondefective	1188
Class A	91
Class B	47
Class C	10

Does this lot differ from previous experience with regard to proportions of thermometers in each category?

The expected numbers in the four categories are 1162.32, 120.24, 40.08, 13.36. This yields a value $\chi^2 = 9.72$ which is significant at the 5% level. The new lot is different from the previous lots.

9.5 Problems

1. The sleep of ten patients was measured after Treatment (1) with dextro-hyoscyamine hydrobromide, and Treatment (2) with

levo-hyoscyamine hydrobromide. The average number of hours of sleep gained or lost is given in Table 9.3.†

Test the hypothesis that Treatment (2) is more effective than Treatment (1).

Table 9.3

Hours of Sleep Gained or Lost under Treatment (1) or Treatment (2)

Patient	Treatment (1)	Treatment (2)
1	0.7	1.9
2	−1.6	0.8
3	−0.2	1.1
4	−1.2	0.1
5	−0.1	−0.1
6	3.4	4.4
7	3.7	5.5
8	0.8	1.6
9	0	4.6
10	2.0	3.4

2. Use the data of Problem 1 to give a 5% and a 1% confidence interval for the difference of the means of the treatments.
3. Give a test for the effectiveness of Treatment (1) and also of Treatment (2).
4. Use the data of Example 8.3.1 to test the hypothesis that there is no difference between Methods A and B.
5.‡ The data given in Table 9.4 are measurements of the capacity of paired batteries, one from each of two different manufacturers.
    (*a*) Test whether there is a difference in the performance of the two sets of batteries.
    (*b*) State the assumptions on which the test is based.

---

† Data from [3, p. 30].
‡ From [2, p. 3.32].

9.5 PROBLEMS

Table 9.4

Measurements of Battery Capacity[a]

A	B
146	141
141	143
135	139
142	139
140	140
143	141
138	138
137	140
142	142
137	138

[a] Values given in ampere hours.

6.[†] There are equal numbers of four different types of meters in service and it is known that all types are equally likely to break down. The actual number of breakdowns is given in the following list:

Type of meter	Number of breakdowns
1	30
2	40
3	33
4	47

Do we have evidence to conclude that the chances of failure for the four types are not equal?

---

† From [2, p. 9.4].

7. A new medicine against hypertension was tested on 18 patients. After 40 days of treatment the following changes of the diastolic blood pressure were observed:

-5, -1, $+2$, $+8$, -25, $+1$, $+5$, -12, -16
-9, -8, -18, -5, -22, $+4$, -21, -15, -11

Is the treatment effective at the 5% level of significance?

8. Twelve patients having high albumin content in their blood were treated with a medicine. Their blood content of albumin was measured before and after treatment. The measured values are given in Table 9.5.

Table 9.5

Blood Content of Albumin[a]

Patient N	Before treatment	After treatment
1	5.02	4.66
2	5.08	5.15
3	4.75	4.30
4	5.25	5.07
5	4.80	5.38
6	5.77	5.10
7	4.85	4.80
8	5.09	4.91
9	6.05	5.22
10	4.77	4.50
11	4.85	4.85
12	5.24	4.56

[a] Values given in grams per 100 ml.

Is the effect of the medicine significant at the 5% level?

9. During a 24-hr period the urine of 18 healthy persons contained, on the average, 20.8 mg tyrosine with a standard deviation of 1.7 mg. 11 patients, suffering from a nutritive disease, had an average tyrosine content of only 13.3 mg with a standard deviation of 2.8 mg. Is the difference significant at the 5% level?

9.5 PROBLEMS

10. The neutrino radiation from outer space was observed during several days. The frequencies of signals were recorded for each sidereal hour and are given in Table 9.6.

Table 9.6
Frequency of Neutrino Radiation from Outer Space

Hour starting at	Frequency of signals	Hour starting at	Frequency of signals
0	24	12	29
1	24	13	26
2	36	14	38
3	32	15	26
4	33	16	37
5	36	17	28
6	41	18	43
7	24	19	30
8	37	20	40
9	37	21	22
10	49	22	30
11	51	23	42

Test whether the signals are uniformly distributed over the 24 hour period. (Test at the 1% level of significance.)

11. Neutrino radiation was observed over a certain period and the number of hours in which 0, 1, 2, ... signals were received was recorded.

Number of signals per hour	Number of hours with this frequency of signals
0	1924
1	541
2	103
3	17
4	1
5	1
6 or more	0

Test whether the observations come from a population which has a Poisson distribution with expected value 0.3.

12.† During World War II, bacterial polysaccharides were investigated as blood plasma extenders. Sixteen samples of hydrolized polysaccharides were supplied by various manufacturers in order to assess two chemical methods for determining their average molecular weight.

Method A	Method B
62,700	56,400
29,100	27,500
44,400	42,200
47,800	46,800
36,300	33,300
40,000	37,100
43,400	37,300
35,800	36,200
33,900	35,200
44,200	38,000
34,300	32,200
31,300	27,300
38,400	36,100
47,100	43,100
42,100	38,400
42,200	39,900

Test whether the average of Method A exceeds the average of Method B.

13. Consider a sample of size n from a normal population with unknown mean θ and known standard deviation $\sigma = 1$. Use the region $\overline{X} > \dfrac{1.645}{\sqrt{n}}$ for testing the hypothesis $\theta = 0$.

(a) Show that the size of this critical region is 5%.

(b) Construct at least two different critical regions of size 2%.

---

† From [2, p. 3.39].

14. Compute the power function of the critical region of Problem 13(a) for $n = 4$ and compare its power with the power of the region used in Example 9.2.1.

15. Consider a family of rectangular populations, that is populations with frequency functions

$$f(x|\theta) = \begin{cases} \dfrac{1}{\theta} & \text{if } 0 \leq x \leq \theta, \\ 0 & \text{otherwise,} \end{cases}$$

where $\theta > 0$ is a parameter. Take a sample of size $n = 2$ to test the hypothesis $H_0 : \theta = \theta_0$, and use as the critical region W_0 the region where at least one of the two observations falls into either the interval $\left(0, \dfrac{a}{2}\right)$ or the interval $\left(\theta_0 - \dfrac{a}{2}, \theta_0\right)$.

   (a) Determine a so as to make the size of the region equal to α.
   (b) Compute the power function of the region W_0 of size α.

16. A coin is tossed 50 times, heads appears 28 times, tails 22 times. Test the hypothesis that the coin is unbiased.

17. A die is tossed 120 times, the faces 1, 2, 3, 4, 5, and 6 appear 16, 30, 32, 18, 11, and 13 times, respectively. Test whether the die is loaded.

References

1. H. Cramér, "Mathematical Methods of Statistics," Princeton Univ. Press, Princeton, New Jersey, 1946.
2. M. Natrella, Experimental Statistics (NBS Handbook 99) U.S. Govt. Printing Office, Washington, D.C., 1963.
3. Student, The probable error of the mean. *Biometrika* **6**, 1–25 (1908). This paper is also reproduced in Student's "Collected Papers," p. 30, Cambridge Univ. Press, London and New York, 1958.

A | Some Combinatorial Formulas

One often must determine the number of ways in which n distinguishable objects can be ordered. Clearly, we have n possible choices for the first position, $(n-1)$ for the second, and so on until we finally have only one choice for the last item. The total number of possible orderings is therefore

$$n(n-1)\cdots 2 \cdot 1.$$

This number is called the number of permutations of n elements (items). It is customary to write

$$n! = 1 \cdot 2 \cdots n \qquad (A.1)$$

for the product of the first n positive integers and to read the symbol $n!$ as n factorial.

The number of permutations is only defined for integers $n > 0$. It is convenient also to define $0!$ to be equal to 1.

Example A.1

We list all possible permutations of the first four natural numbers

1234	2134	3124	4123
1243	2143	3142	4132
1324	2314	3214	4213
1342	2341	3241	4231
1423	2413	3412	4312
1432	2431	3421	4321

We see that $24 = 4! = 1 \cdot 2 \cdot 3 \cdot 4$ permutations are possible.

A second problem which is often important is the determination of the number of ways in which k objects can be selected from n distinct objects, without regard to the order in which the objects are selected.

We first determine the number of ways in which k of the n items can be selected taking into account their order. One has again n possibilities for choosing the first item, $(n - 1)$ for the second, and so on to $(n - k + 1)$ for the kth item. The total number of ways of selecting k ordered items is then

$$n(n-1)\cdots(n-k+1) = \frac{n!}{(n-k)!} \qquad (A.2)$$

Example A.2

We select two out of the first four integers, taking into account their order

| 12 | 13 | 14 | 23 | 24 | 34 |
| 21 | 31 | 41 | 32 | 42 | 43 |

We obtain $\frac{4!}{2!} = \frac{24}{2} = 12$ possible selections.

If we wish to disregard the ordering of the items we selected, we

SOME COMBINATORIAL FORMULAS

must only count once those of the $\dfrac{n!}{(n-k)!}$ arrangements which differ only in their ordering. The k items selected admit $k!$ orderings, the total number of selections—disregarding the order of the selected items—is therefore

$$\frac{n!}{k!\,(n-k)!}.$$

We write

$$\binom{n}{k} = \frac{n!}{k!\,(n-k)!}; \qquad (A.3)$$

the symbol $\binom{n}{k}$ is called a "binomial coefficient." The symbol $\binom{n}{k}$ is originally defined only for $k = 1, 2, \ldots, n-1$. It is, however, convenient to define also

$$\binom{n}{0} = 1 \quad \text{and} \quad \binom{n}{n} = 1.$$

This is in accordance with our agreement to put $0! = 1$. The binomial coefficient $\binom{n}{k}$ is the number of combinations of n elements in groups of k.

Example A.3

If we select three out of the first five positive integers, disregarding the order of the numbers selected we obtain

```
123    234    345
124    235
125    245
134
135
145
```

There are $\binom{5}{3} = \frac{5 \cdot 4 \cdot 3}{1 \cdot 2 \cdot 3}$ combinations of five elements in groups of three.

The binomial coefficients satisfy the following important relations:

$$\binom{n}{k} = \binom{n}{n-k}; \qquad (A.4)$$

$$\binom{n+1}{k} = \binom{n}{k} + \binom{n}{k-1}. \qquad (A.5)$$

Equation (A.4) follows immediately from (A.3). To prove (A.5) we use (A.3) and obtain

$$\binom{n}{k} + \binom{n}{k-1} = \frac{n!}{k!(n-k)!} + \frac{n!}{(k-1)!(n-k+1)!}$$

$$= \frac{n!}{k!(n-k+1)!}[n-k+1+k]$$

$$= \binom{n+1}{k}.$$

Formula (A.5) can be used to derive the following result.

Binomial Theorem

$$(a+b)^n = \sum_{k=0}^{n} \binom{n}{k} a^k b^{n-k}.$$

We prove the binomial theorem by induction. Its statement is true for $n = 1$. We must therefore assume that it is valid for n and show that this assumption implies that the formula holds also for $n + 1$. We have

$$(a+b)^{n+1} = (a+b)(a+b)^n$$

$$= (a+b) \sum_{k=0}^{n} \binom{n}{k} a^k b^{n-k}$$

$$= \sum_{k=0}^{n} \binom{n}{k} a^{k+1} b^{n-k} + \sum_{k=0}^{n} \binom{n}{k} a^k b^{n-k+1}.$$

SOME COMBINATORIAL FORMULAS

We change the index of summation in the first sum by putting $k = j - 1$ and detach the last term from the first sum and the first term from the last sum. In this way we obtain

$$(a+b)^{n+1} = a^{n+1} + \sum_{j=1}^{n}\left[\binom{n}{j-1} + \binom{n}{j}\right]a^j b^{n+1-j} + b^{n+1}.$$

We finally use (A.5) to complete the proof.

The computation of factorials for large values of n is cumbersome and often inconvenient. Tools of higher analysis permit the derivation of an excellent approximation, which we mention without proof.

$$n! = \left(\frac{n}{e}\right)^n \sqrt{2\pi n}\, e^{r(n)}, \tag{A.6}$$

where $0 \leq r(n) \leq 1$ and where $r(n)$ tends to zero as n tends to infinity. Formula (A.6) is called Stirling's formula.

We use Stirling's formula to derive a useful approximation for the binomial coefficients $\binom{n}{k}$ valid if n, k, and $n-k$ tend to infinity. We have

$$\binom{n}{k} = \frac{n!}{k!(n-k)!}$$

$$= \frac{\left(\frac{n}{e}\right)^n \sqrt{2\pi n}}{\left(\frac{k}{e}\right)^k \sqrt{2\pi k}\left(\frac{n-k}{e}\right)^{n-k}\sqrt{2\pi(n-k)}} \exp[r(n) - r(k) - r(n-k)]$$

or putting $R = r(n) - r(k) - r(n-k)$

$$\binom{n}{k} = \frac{n^{n+(1/2)} e^R}{k^{k+(1/2)}(n-k)^{n-k+(1/2)}\sqrt{2\pi}}.$$

A simple computation yields then the formula

$$\binom{n}{k} = \frac{1}{\sqrt{2\pi n}}\left(\frac{n}{k}\right)^{k+(1/2)}\left(\frac{n}{n-k}\right)^{n-k+(1/2)} e^R. \tag{A.7}$$

Since R tends to zero as n, k, and $n - k$ tend to infinity, we see that the quotient

$$\frac{\binom{n}{k}}{\frac{1}{\sqrt{2\pi n}} \left(\frac{n}{k}\right)^{k+(1/2)} \left(\frac{n}{n-k}\right)^{n-k+(1/2)}} \tag{A.8}$$

is approximately equal to 1 if n, k, and $n-k$ are sufficiently large. The denominator of (A.8) can therefore be used as an approximation of the binomial coefficient $\binom{n}{k}$.

B | The Gamma Function

It is known† that the integral $\int_0^\infty t^{x-1}e^{-t}\,dt$ converges for all $x > 0$. We can therefore define the function

$$\Gamma(x) = \int_0^\infty t^{x-1}e^{-t}\,dt \tag{B.1}$$

for all x such that $0 < x < \infty$. The function $\Gamma(x)$ is called the gamma function. It has the following important property:

$$\Gamma(x+1) = x\Gamma(x) \qquad (0 < x < \infty). \tag{B.2}$$

---

† See D. V. Widder, "Advanced Calculus," 2nd ed. Prentice-Hall, Englewood Cliffs, New Jersey, 1960.

We obtain relation (B.2) by integrating $\Gamma(x + 1) = \int_0^\infty t^x e^{-t}\, dt$ by parts. Iterating (B.2) we see that for $x > 0$ and a positive integer n

$$\Gamma(x + n) = (x + n - 1)(x + n - 2) \cdots x\Gamma(x). \tag{B.3}$$

We also see that

$$\Gamma(1) = \int_0^\infty e^{-t}\, dt = 1. \tag{B.4}$$

Let n be a positive integer, we see then from (B.2) and (B.4) that

$$\Gamma(n) = (n - 1)!. \tag{B.5}$$

It is easy to prove that the gamma function is continuous for $x > 0$. In view of (B.5) the function $\Gamma(x)$ is a solution of the problem of interpolating between the factorials by means of a continuous function.

It can also be shown that

$$\Gamma(\tfrac{1}{2}) = \sqrt{\pi}. \tag{B.6}$$

The knowledge of this value of the gamma function is often important, it can, for instance, be used to derive the relation (4.1.3).

C | Proof of the Central Limit Theorem

In this appendix, we give the proof of a theorem that is somewhat more general than Theorem 5.2.3 and that contains Theorem 5.2.3 as a particular case. In order to avoid analytical methods requiring many prerequisites, we apply a technique that can easily be presented in an elementary way. This technique consists in adjoining to each distribution function an operator useful in proving limit theorems.

C.1 Operators

Let C_3 be the set of all uniformly bounded and uniformly continuous functions defined on the real line R_1 which can be differentiated three times and whose first, second, and third derivatives are also uniformly bounded and uniformly continuous on R_1.

Let $f \in C_3$, we write then

$$\|f\| = \sup_x |f(x)| \qquad \text{(C.1.1)}$$

and call $\|f\|$ the norm of the function $f(x)$.

It is easily seen that the relations $f \in C_3$, $g \in C_3$, and α real imply that $f + g \in C_3$ and $\alpha f \in C_3$; moreover,

$$\|f + g\| \le \|f\| + \|g\|,$$

and

$$\|\alpha f\| = |\alpha|\, \|f\|.$$

A transformation A, which changes each function $f \in C_3$ into a function $g = g(x) = Af$ of C_3, is called a linear operator if the following three conditions are satisfied:

(i) $A(f + g) = Af + Ag$ if $f, g \in C_3$
(ii) If $f \in C_3$ and if α is real then $A(\alpha f) = \alpha Af$
(iii) There exists a positive number K such that for every $f \in C_3$ the relation $\|Af\| \le K\|f\|$ holds.

An operator is called a contraction operator if (iii) is satisfied for $K = 1$.

We define the sum $A + B$ and the product AB of two operators A and B by the relations

$$(A + B)f = Af + Bf,$$
$$(AB)f = A(Bf).$$

The addition of operators is commutative and associative, whereas the multiplication of operators is associative but, in general, not commutative. The distribution law

$$A(B + C) = AB + AC$$

is also valid. If α and β are real numbers then the operators αA, $\alpha(A + B)$, $\alpha(\beta A)$, $(\alpha + \beta)A$ are defined by $(\alpha A)f = \alpha Af$, $\alpha(A + B)f = \alpha Af + \alpha Bf$,

$\alpha(\beta A)f = (\alpha\beta)Af$, $(\alpha + \beta)Af = \alpha Af + \beta Af$. Let 0 be the zero operator, that is, an operator that transforms each $f \in C_3$ into the function which is identically zero, then $A + (-A) = 0$.

C.2 Convolutions and the Addition of Independent Random Variables

In this section we discuss some topics that supplement Section 6.1.

Theorem C.2.1A

Let $F(x)$ and $G(x)$ be two absolutely continuous distribution functions then

$$H(x) = \int_{-\infty}^{\infty} F(x-y)G'(y)\,dy = \int_{-\infty}^{\infty} G(x-y)F'(y)\,dy. \quad \text{(C.2.1a)}$$

is also an absolutely continuous distribution and its frequency function is given by

$$H'(x) = \int_{-\infty}^{\infty} F'(x-y)G'(y)\,dy = \int_{-\infty}^{\infty} G'(x-y)F'(y)\,dy. \quad \text{(C.2.1b)}$$

The statement follows immediately from the definition of a distribution function and is closely related to Theorem 6.1.3.

In the same way we obtain

Theorem C.2.1B

Let $F(x) = \sum_k p_k \varepsilon(x - \xi_k)$ and $G(x) = \sum_j q_j \varepsilon(x - \eta_j)$ be two discrete distributions then

$$H(x) = \sum_k \sum_j p_k q_j \varepsilon(x - \xi_k - \eta_j) \quad \text{(C.2.1c)}$$

is also a discrete distribution function.

Remark 1

The distribution function $H(x)$ introduced by formulas (C.2.1a) and (C.2.1c) is called the convolution of F and G and one writes frequently $H = F * G$.

Remark 2

It follows from (C.2.1a) and (C.2.1c) that convolutions are commutative, that is $H = F * G = G * F$. It is also easily seen that convolutions are associative, that is, if F_1, F_2 and F_3 are distribution functions then $(F_1 * F_2) * F_3 = F_1 * (F_2 * F_3)$.

Theorem C.2.2

Let X and Y be two independent random variables and denote their distribution functions by $F(x)$ and $G(y)$, respectively. The distribution function $H(x)$ of the sum $S = X + Y$ is $H = F * G$, the convolution of the distribution functions of the summands.

Remark 1

Theorem C.2.2 refers to arbitrary random variables. The proof of the case where F and G are both absolutely continuous is given in the main body of the text (Theorem 6.1.3). The method of proving Theorem 6.1.3 can easily be modified so as to apply also if F and G are both discrete.

Remark 2

Theorem C.2.2 can be extended (by induction) to the sum of an arbitrary number of random variables. Let $X_1, X_2, \ldots, X_n$ be n independent random variables with distribution functions $F_1, F_2, \ldots, F_n$, respectively. The distribution function of the sum $S = X_1 + X_2 + \cdots + X_n$ is then the convolution $F_1 * F_2 * \cdots * F_n$.

C.3 Probability Operators

Let $F(x)$ be an absolutely continuous distribution function, we adjoin to it an operator A_F by defining for all $f \in C_3$

PROOF OF THE CENTRAL LIMIT THEOREM

$$A_F f = \int_{-\infty}^{\infty} f(x + y) F'(y) \, dy \quad \text{(C.3.1a)}$$

Let $F(x)$ be a discrete distribution function, $F(x) = \sum_j p_j \varepsilon(x - \xi_j)$; in this case we define

$$A_F f = \sum_j p_j f(x + \xi_j). \quad \text{(C.3.1b)}$$

The operator A_F is called the probability operator belonging to the distribution function $F(x)$. It follows from (C.3.1a) that $A_F f \in C_3$.

Remark

In this appendix, we treat only absolutely continuous and discrete distributions, the method of probability operators can, however, be applied to arbitrary distributions.

Theorem C.3.1

The probability operator A_F of an arbitrary distribution function F is a linear contraction operator.

It is obvious that Conditions (*i*) and (*ii*) of Section C.1 are satisfied, moreover,

$$\|A_F f\| \le \|f\| \int_{-\infty}^{\infty} F'(x) \, dx = \|f\|$$

in case F is absolutely continuous, while

$$\|A_F f\| \le \|f\| \sum_j p_j = \|f\|$$

if F is discrete. Therefore, A_F is a contraction operator.

Theorem C.3.2

Let F and G be two arbitrary distribution functions. Then their operators A_F and A_G commute and $A_F A_G = A_H$ where $H = F * G$.

Proof

We give the proof only for the case where F and G are both absolutely continuous. Let $f \in C_3$, then

$$A_F A_G f = \int_{-\infty}^{\infty} \left[\int_{-\infty}^{\infty} f(x + y + z) G'(z) \, dz \right] F'(y) \, dy$$

$$= \int_{-\infty}^{\infty} f(x + u) H'(u) \, du.$$

The fact that the operators A_F and A_G are commutative follows immediately from the commutativity of the convolution.

The case of two discrete distributions is treated in a similar way.

If we combine Theorem 6.1.3 with Theorem 6.3.2, we obtain the following important result:

Corollary to Theorem C.3.2

Let $X_1, X_2, \ldots, X_n$ be n independent random variables, and denote the distribution function of X_j by F_j ($j = 1, 2, \ldots, n$). Let $S_n = \sum_{j=1}^{n} X_j$ and denote the distribution function of S_n by G. Then $G = F_1 * F_2 * \cdots * F_n$ and the probability operator of G is given by $A_G = A_{F_1} A_{F_2} \cdots A_{F_n}$, where A_{F_j} is the probability operator that belongs to F_j.

Lemma C.3.1

Let A_F and A_G be the operators that belong to the distribution functions F and G, respectively, and let $f \in C_3$, then $\|A_F A_G f\| \leq \|A_G f\|$.

The statement follows from the fact that A_F is a contraction operator.

Lemma C.3.2

Let $U_1, U_2, \ldots, U_n$ and $V_1, V_2, \ldots, V_n$ be probability operators that belong to certain distribution functions, and let $f \in C_3$, then

$$\| U_1 U_2 \cdots U_n f - V_1 V_2 \cdots V_n f \| \leq \sum_{k=1}^{n} \| U_k f - V_k f \|.$$

PROOF OF THE CENTRAL LIMIT THEOREM

The relation

$$U_1 U_2 \cdots U_n - V_1 V_2 \cdots V_n = \sum_{k=1}^{n} U_1 \cdots U_{k-1}(U_k - V_k) V_{k+1} \cdots V_n$$

is valid for arbitrary linear operators and can easily be verified. If the U_j and V_j ($j = 1, 2, \ldots, n$) are probability operators, then we conclude from Theorem C.3.1, Theorem C.3.2, and Lemma C.3.1 the statement of Lemma C.3.2.

C.4 The Limit Theorem

We prove here a theorem that permits the derivation of Theorem 5.2.3.

Theorem C.4.1

Let $\{X_k\}$ be a sequence of independent random variables, and denote the distribution function of X_k by $F_k(x)$. Suppose that $F_k(x)$ is absolutely continuous and that $\mathscr{E}(X_k) = \alpha_k$ and $\mathrm{Var}(X_k) = \sigma_k^2$ exist. Let $s_n^2 = \sum_{k=1}^{n} \sigma_k^2$ and suppose that the relation

$$\lim_{n \to \infty} \frac{1}{s_n^2} \sum_{k=1}^{n} \int_{|x| > \varepsilon s_n} x^2 F_k'(x + \alpha_k) \, dx = 0 \qquad (\text{C.4.1})$$

is satisfied for all $\varepsilon > 0$. Let $S_n^* = \frac{1}{s_n} \sum_{k=1}^{n} (X_k - \alpha_k)$, and denote the distribution function of S_n^* by F_n^*. Then

$$\lim_{n \to \infty} F_n^*(x) = \Phi(x) = \frac{1}{\sqrt{2\pi}} \int_{-\infty}^{x} \exp\left(-\frac{y^2}{2}\right) dy.$$

Remark 1

The assumption that the $F_k(x)$ are absolutely continuous is unnecessary. It is only made to avoid the use of more advanced analytical tools.

Remark 2

Theorem C.4.1 differs from Theorem 5.2.3 in two respects:

(*a*) in Theorem 5.2.3 we assume that the X_j are identically distributed;
(*b*) In Theorem 5.2.3 we do not need condition (C.4.1).

Remark 3

Condition (C.4.1) is usually called the Lindeberg condition.

In proving Theorem C.4.1 it is no restriction to assume that $\alpha_k = 0$ and $s_n = \left[\sum_{k=1}^{n} \mathrm{Var}(X_k)\right]^{1/2} = 1$. This is easily seen by introducing the new variables $Y_k = \dfrac{X_k - \alpha_k}{s_n}$. In proving the theorem we assume therefore that $s_n = 1$ and that the Lindeberg condition has the form

$$\lim_{n \to \infty} \sum_{k=1}^{n} \int_{|x| > \varepsilon} x^2 F'_k(x)\, dx = 0. \qquad \text{(C.4.1a)}$$

Let U_k be the probability operator of F_k and let V_k be the probability operator that belongs to a normal distribution with mean zero and variance σ_k^2. Since we assumed that $\alpha_k = 0$ and $s_n = 1$, we see that $S_n^* = \sum_{j=1}^{n} X_j$. It follows then from the remarks to Theorem C.2.2 and from the Corollary to Theorem C.3.2 that the operator A_n of the distribution function F_n^* is

$$A_n = U_1 U_2 \cdots U_n.$$

Similarly, we conclude from the relation

$$\Phi(x) = \Phi\left(\frac{x}{\sigma_1}\right) * \Phi\left(\frac{x}{\sigma_2}\right) * \cdots * \Phi\left(\frac{x}{\sigma_n}\right)$$

that the operator B of Φ is given by

$$B = V_1 V_2 \cdots V_k.$$

PROOF OF THE CENTRAL LIMIT THEOREM

It follows from Lemma C.3.2 that for $f \in C_3$

$$\|A_n f - Bf\| \leq \sum_{k=1}^{n} \|U_k f - V_k f\|. \tag{C.4.2}$$

Our next aim is to show that

$$\lim_{n \to \infty} \|A_n f - Bf\| = 0. \tag{C.4.3}$$

In order to determine $U_k f$ we expand $f(x + y)$ according to Taylor's formula:

$$f(x + y) = f(x) + yf'(x) + \frac{y^2}{2} f''(x + \theta_1 y), \tag{C.4.4}$$

$$f(x + y) = f(x) + yf'(x) + \frac{y^2}{2} f''(x) + \frac{y^3}{6} f'''(x + \theta_2 y), \tag{C.4.5}$$

and note that for an arbitrary $\varepsilon > 0$

$$U_k f = \int_{-\varepsilon}^{\varepsilon} f(x + y) F_k'(y) \, dy + \int_{|y| > \varepsilon} f(x + y) F_k'(y) \, dy.$$

We substitute (C.4.5) into the first and (C.4.4) into the second integral and obtain

$$U_k f = f(x) + \tfrac{1}{2} f''(x) \sigma_k^2 + \tfrac{1}{6} \int_{|y| \leq \varepsilon} y^3 f'''(x + \theta_2 y) F_k'(y) \, dy$$

$$+ \tfrac{1}{2} \int_{|y| > \varepsilon} y^2 [f''(x + \theta_1 y) - f''(x)] F_k'(y) \, dy. \tag{C.4.6}$$

Let $M_2 = \sup |f''(x)|$ and $M_3 = \sup |f'''(x)|$, it follows then from (C.4.6) that

$$\left| U_k f - f(x) - \frac{1}{2} \sigma_k^2 f''(x) \right| \leq \frac{\varepsilon M_3}{6} \sigma_k^2 + M_2 \int_{|y| > \varepsilon} y^2 F_k'(y) \, dy. \tag{C.4.7a}$$

We also have

$$V_k f = \frac{1}{\sigma_k \sqrt{2\pi}} \int_{-\infty}^{\infty} f(x + y) \exp\left(-\frac{y^2}{2\sigma_k^2}\right) dy,$$

using the expansion (C.4.5) we obtain

$$\left| V_k f - f(x) - \frac{\sigma_k^2}{2} f''(x) \right| \leq \frac{M_3}{3} \sigma_k^3. \qquad \text{(C.4.7b)}$$

It follows from (C.4.7a) and (C.4.7b) that

$$\|U_k f - V_k f\| \leq \varepsilon \frac{M_3}{6} \sigma_k^2 + M_2 \int_{|y|>\varepsilon} y^2 F_k'(y)\,dy + \frac{M_3}{3} \sigma_k^3;$$

since $\sum_{k=1}^{n} \sigma_k^2 = 1$, we see that

$$\sum_{k=1}^{n} \|U_k f - V_k f\| \leq \frac{\varepsilon}{6} M_3 + M_2 \sum_{k=1}^{n} \int_{|y|>\varepsilon} y^2 F_k'(y)\,dy + \frac{1}{3} M_3 \sum_{k=1}^{n} \sigma_k^3.$$

$$\text{(C.4.8)}$$

We have

$$\sigma_k^2 = \int_{|x|\leq\varepsilon} y^2 F_k'(y)\,dy + \int_{|x|>\varepsilon} y^2 F_k'(y)\,dy \leq \varepsilon^2 + \int_{|x|>\varepsilon} y^2 F_k'(y)\,dy.$$

Since—by assumption—$\sum_{k=1}^{n} \sigma_k^2 = 1$, we see from the last inequality that

$$\sum_{k=1}^{n} \sigma_k^3 \leq \max_{1\leq k\leq n} \sigma_k \leq \left[\varepsilon^2 + \sum_{k=1}^{n} \int_{|x|>\varepsilon} y^2 F_k'(y)\,dy \right]^{1/2} \qquad \text{(C.4.9)}$$

and we conclude from (C.4.8), (C.4.9) and the Lindeberg condition (C.4.1a) that

$$\lim_{n\to\infty} \sum_{k=1}^{n} \|U_k f - V_k f\| = 0.$$

The relation

$$\lim_{n\to\infty} \|A_n f - Bf\| = 0 \qquad \text{[C.4.3]}$$

follows then from (C.4.2). In view of the definition of the norm this means that

$$\lim_{n\to\infty} \int_{-\infty}^{\infty} f(x+y) p_n^*(y)\,dy = \int_{-\infty}^{\infty} f(x+y) \Phi'(y)\,dy \qquad \text{(C.4.10)}$$

for all $f \in C_3$. Here $p_n^*(y) = \dfrac{d}{dy} F_n^*(y)$ is the density function of $F_n^*(y)$.

PROOF OF THE CENTRAL LIMIT THEOREM

Let $\varepsilon > 0$ be an arbitrary positive number and consider a function $f_\varepsilon(x) \in C_3$ which has the following properties: $f_\varepsilon(x) = 1$ for $x \leq 0$, $f_\varepsilon(x) = 0$ for $x \geq \varepsilon$, while $f_\varepsilon(x)$ is monotone decreasing in the interval $(0, \varepsilon)$. The existence of such functions is assured by the following example:

$$f_\varepsilon(x) = \begin{cases} 1 & \text{if } x \leq 0, \\ \left[1 - \left(\frac{x}{\varepsilon}\right)^4\right]^4 & \text{if } 0 \leq x \leq \varepsilon, \\ 0 & \text{if } \varepsilon \leq x. \end{cases}$$

From the definition of the function $f_\varepsilon(x)$ one easily concludes that

$$\Phi(-x + \varepsilon) \geq \int_{-\infty}^{\infty} f_\varepsilon(x + y)\Phi'(y)\,dy \geq \Phi(-x)$$

for all real x. Similarly we see that

$$F_n^*(-x + \varepsilon) \geq \int_{-\infty}^{\infty} f_\varepsilon(x + y)p_n^*(y)\,dy \geq F_n^*(-x)$$

for all real x. It follows from the last two relations and from (C.4.10) that

$$\limsup_{n \to \infty} F_n^*(x) \leq \Phi(x + \varepsilon) \tag{C.4.11}$$

and

$$\liminf_{n \to \infty} F_n^*(x + \varepsilon) \geq \Phi(x)$$

for all real x and arbitrary $\varepsilon > 0$. The last inequality can be written as

$$\liminf_{n \to \infty} F_n^*(x) \geq \Phi(x - \varepsilon). \tag{C.4.12}$$

We combine (C.4.11) and (C.4.12) and see that

$$\Phi(x - \varepsilon) \leq \liminf_{n \to \infty} F_n^*(x) \leq \limsup_{n \to \infty} F_n^*(x) \leq \Phi(x + \varepsilon)$$

for all real x and arbitrary $\varepsilon > 0$. The statement of Theorem C.4.1 follows immediately.

In order to derive Theorem 5.2.1 as a particular case, we assume that the random variables $\{X_k\}$ are identically distributed and have finite variance. Theorem 5.2.3 is then proven as soon as one shows that the Lindeberg condition is satisfied. Let $F(x)$ be the common distribution function of the X_k and let

$$\alpha = \int_{-\infty}^{\infty} xF'(x)\,dx \quad \text{and} \quad \sigma^2 = \int_{-\infty}^{\infty} x^2 F'(x+\alpha)\,dx$$

be the mean and the variance, respectively, of $F(x)$. Then

$$s_n^2 = n\sigma^2,$$

and

$$\frac{1}{s_n^2} \sum_{k=1}^{n} \int_{|x| > \varepsilon s_n} x^2 F_k'(x + \alpha_k)\,dx$$

becomes

$$\frac{1}{\sigma^2} \int_{|x| > \varepsilon \sigma \sqrt{n}} x^2 F'(x + \alpha)\,dx.$$

Since we assumed the existence of the variance we see that this expression tends to zero as $n \to \infty$. The Lindeberg condition is therefore satisfied if the X_k are identically distributed, so that Theorem 5.2.3 is proven.

References

H. F. Trotter, An elementary proof of the central limit theorem. *Arch. Math.* **10**, 226–234 (1959).

W. Feller, "An Introduction to Probability Theory and its Applications," Vol. II. Wiley, New York, 1966.

D Tables

Table I. Normal Distribution[a]

$$\Phi(x) = \int_{-\infty}^{x} \frac{1}{\sqrt{2\pi}} \exp\left(-\frac{t^2}{2}\right) dt$$

x	.00	.01	.02	.03	.04	.05	.06	.07	.08	.09
.0	.5000	.5040	.5080	.5120	.5160	.5199	.5239	.5279	.5319	.5359
.1	.5398	.5438	.5478	.5517	.5557	.5596	.5636	.5675	.5714	.5753
.2	.5793	.5832	.5871	.5910	.5948	.5987	.6026	.6064	.6103	.6141
.3	.6179	.6217	.6255	.6293	.6331	.6368	.6406	.6443	.6480	.6517
.4	.6554	.6591	.6628	.6664	.6700	.6736	.6772	.6808	.6844	.6879
.5	.6915	.6950	.6985	.7019	.7054	.7088	.7123	.7157	.7190	.7224
.6	.7257	.7291	.7324	.7357	.7389	.7422	.7454	.7486	.7517	.7549
.7	.7580	.7611	.7642	.7673	.7704	.7734	.7764	.7794	.7823	.7852
.8	.7881	.7910	.7939	.7967	.7995	.8023	.8051	.8078	.8106	.8133
.9	.8159	.8186	.8212	.8238	.8264	.8289	.8315	.8340	.8365	.8389
1.0	.8413	.8438	.8461	.8485	.8508	.8531	.8554	.8577	.8599	.8621
1.1	.8643	.8665	.8686	.8708	.8729	.8749	.8770	.8790	.8810	.8830
1.2	.8849	.8869	.8888	.8907	.8925	.8944	.8962	.8980	.8997	.9015
1.3	.9032	.9049	.9066	.9082	.9099	.9115	.9131	.9147	.9162	.9177
1.4	.9192	.9207	.9222	.9236	.9251	.9265	.9279	.9292	.9306	.9319
1.5	.9332	.9345	.9357	.9370	.9382	.9394	.9406	.9418	.9429	.9441
1.6	.9452	.9463	.9474	.9484	.9495	.9505	.9515	.9525	.9535	.9545
1.7	.9554	.9564	.9573	9582	9591	.9599	.9608	.9616	.9625	.9633
1.8	.9641	.9649	.9656	.9664	.9671	.9678	.9686	.9693	.9699	.9706
1.9	.9713	.9719	.9726	.9732	.9738	.9744	.9750	.9756	.9761	.9767
2.0	.9772	.9778	.9783	.9788	.9793	.9798	.9803	.9808	.9812	.9817
2.1	.9821	.9826	.9830	.9834	.9838	.9842	.9846	.9850	.9854	.9857
2.2	.9861	.9864	.9868	.9871	.9875	.9878	.9881	.9884	.9887	.9890
2.3	.9893	.9896	.9898	.9901	.9904	.9906	.9909	.9911	.9913	.9916
2.4	.9918	.9920	.9922	.9925	.9927	.9929	.9931	.9932	.9934	.9936
2.5	.9938	.9940	.9941	.9943	.9945	.9946	.9948	.9949	.9951	.9952
2.6	.9953	.9955	.9956	.9957	.9959	.9960	.9961	.9962	.9963	.9964
2.7	.9965	.9966	.9967	.9968	.9969	.9970	.9971	.9972	.9973	.9974
2.8	.9974	.9975	.9976	.9977	.9977	.9978	.9979	.9979	.9980	.9981
2.9	.9981	.9982	.9982	.9983	.9984	.9984	.9985	.9985	.9986	.9986
3.0	.9987	.9987	.9987	.9988	.9988	.9989	.9989	.9989	.9990	.9990
3.1	.9990	.9991	.9991	.9991	.9992	.9992	.9992	.9992	.9993	.9993
3.2	.9993	.9993	.9994	.9994	.9994	.9994	.9994	.9995	.9995	.9995
3.3	.9995	.9995	.9995	.9996	.9996	.9996	.9996	.9996	.9996	.9997
3.4	.9997	.9997	.9997	.9997	.9997	.9997	.9997	.9997	.9997	.9998

[a] From A. M. Mood "Introduction to the Theory of Statistics," copyright McGraw-Hill 1950, 1963, used with permission of McGraw-Hill Book Company.

Table IA

The Normal Distribution[a,b]

p	λ_p
.20	1.282
.10	1.645
.05	1.960
.02	2.326
.01	2.576
.002	3.090
.001	3.291
.0001	3.891
.00001	4.417

[a] The $p\%$ values λ_p of the normal distribution with mean zero and variance one are defined by $2[1 - \Phi(\lambda_p)] = \dfrac{p}{100}$.

[b] From A. M. Mood, "Introduction to the Theory of Statistics," copyright McGraw-Hill 1950, 1963, used with permission of McGraw-Hill Book Company.

Table II

The χ^2 Distribution[a,b]

Degrees of freedom n	χ_p^2 as function of n and p								
	$p=99$	98	95	90	20	10	5	2	1	0.1
1	0.000	0.001	0.004	0.016	1.642	2.706	3.841	5.412	6.635	10.827
2	0.020	0.040	0.103	0.211	3.219	4.605	5.991	7.824	9.210	13.815
3	0.115	0.185	0.352	0.584	4.642	6.251	7.815	9.837	11.341	16.268
4	0.297	0.429	0.711	1.064	5.989	7.779	9.488	11.668	13.277	18.465
5	0.554	0.752	1.145	1.610	7.289	9.236	11.070	13.388	15.086	20.517
6	0.872	1.134	1.635	2.204	8.558	10.654	12.592	15.033	16.812	22.457
7	1.230	1.564	2.167	2.833	9.803	12.017	14.067	16.622	18.475	24.322
8	1.646	2.032	2.733	3.490	11.030	13.362	15.507	18.168	20.090	26.125
9	2.088	2.532	3.325	4.168	12.242	14.684	16.919	19.679	21.666	27.877
10	2.558	3.059	3.940	4.865	13.442	15.987	18.307	21.161	23.209	29.588
11	3.053	3.609	4.575	5.578	14.631	17.275	19.675	22.618	24.725	31.264
12	3.571	4.178	5.226	6.304	15.812	18.549	21.026	24.054	26.217	32.909
13	4.107	4.765	5.892	7.042	16.985	19.812	22.362	25.472	27.688	34.528
14	4.660	5.368	6.571	7.790	18.151	21.064	23.685	26.873	29.141	36.123
15	5.229	5.985	7.261	8.547	19.311	22.307	24.996	28.259	30.578	37.697
16	5.812	6.614	7.962	9.312	20.465	23.542	26.296	29.633	32.000	39.252
17	6.408	7.255	8.672	10.085	21.615	24.769	27.587	30.995	33.409	40.790
18	7.015	7.906	9.390	10.865	22.760	25.989	28.869	32.346	34.805	42.312
19	7.633	8.567	10.117	11.651	23.900	27.204	30.144	33.687	36.191	43.820
20	8.260	9.237	10.851	12.443	25.038	28.412	31.410	35.020	37.566	45.315
21	8.897	9.915	11.591	13.240	26.171	29.615	32.671	36.343	38.932	46.797
22	9.542	10.600	12.338	14.041	27.301	30.813	33.924	37.659	40.289	48.268
23	10.196	11.293	13.091	14.848	28.429	32.007	35.172	38.968	41.638	49.728
24	10.856	11.992	13.848	15.659	29.553	33.196	36.415	40.270	42.980	51.179
25	11.524	12.697	14.611	16.473	30.675	34.382	37.652	41.566	44.314	52.620
26	12.198	13.409	15.379	17.292	31.795	35.563	38.885	42.856	45.642	54.052
27	12.879	14.125	16.151	18.114	32.912	36.741	40.113	44.140	46.963	55.476
28	13.565	14.847	16.928	18.939	34.027	37.916	41.337	45.419	48.278	56.893
29	14.256	15.574	17.708	19.768	35.139	39.087	42.557	46.693	49.588	58.302
30	14.953	16.306	18.493	20.599	36.250	40.256	43.773	47.962	50.892	59.703

[a] For a given percentage p, the $p\%$ value χ_p^2 of the χ^2 distribution is defined by the condition

$$P(\chi^2 > \chi_p^2) = \frac{p}{100}.$$

Thus the probability that χ^2 assumes a value exceeding χ_p^2 is equal to $p\%$.

[b] From H. Cramér, "Elements of Probability Theory." Wiley, New York, 1955. Used with permission of the publisher.

Table III

The t-Distribution[a,b]

Degrees of freedom n	$p=90$	80	60	40	20	10	5	2	1	0.1
				t_p as function of n and p						
1	0.158	0.325	0.727	1.376	3.073	6.314	12.706	31.821	63.657	636.619
2	0.142	0.289	0.617	1.061	1.886	2.920	4.303	6.965	9.925	31.589
3	0.137	0.277	0.584	0.978	1.638	2.353	3.182	4.541	5.841	12.941
4	0.134	0.271	0.569	0.941	1.533	2.132	2.776	3.747	4.604	8.610
5	0.132	0.267	0.559	0.920	1.476	2.015	2.571	3.365	4.032	6.859
6	0.131	0.265	0.553	0.906	1.440	1.943	2.447	3.143	3.707	5.959
7	0.130	0.263	0.549	0.896	1.415	1.895	2.365	2.998	3.499	5.405
8	0.130	0.262	0.546	0.889	1.397	1.860	2.306	2.896	3.355	5.041
9	0.129	0.261	0.543	0.883	1.383	1.833	2.262	2.821	3.250	4.781
10	0.129	0.260	0.542	0.879	1.372	1.812	2.228	2.764	3.169	4.587
11	0.129	0.260	0.540	0.876	1.363	1.796	2.201	2.718	3.106	4.437
12	0.128	0.259	0.539	0.873	1.356	1.782	2.179	2.681	3.055	4.318
13	0.128	0.259	0.538	0.870	1.350	1.771	2.160	2.650	3.012	4.221
14	0.128	0.258	0.537	0.868	1.345	1.761	2.145	2.624	2.977	4.140
15	0.128	0.258	0.536	0.866	1.341	1.753	2.131	2.602	2.947	4.073
16	0.128	0.258	0.535	0.865	1.337	1.746	2.120	2.583	2.921	4.015
17	0.128	0.257	0.534	0.863	1.333	1.740	2.110	2.567	2.898	3.965
18	0.127	0.257	0.534	0.862	1.330	1.734	2.101	2.552	2.878	3.922
19	0.127	0.257	0.533	0.861	1.328	1.729	2.093	2.539	2.861	3.883
20	0.127	0.257	0.533	0.860	1.325	1.725	2.086	2.528	2.845	3.850
21	0.127	0.257	0.532	0.859	1.323	1.721	2.080	2.518	2.831	3.819
22	0.127	0.256	0.532	0.858	1.321	1.717	2.074	2.508	2.819	3.792
23	0.127	0.256	0.532	0.858	1.319	1.714	2.069	2.500	2.807	3.767
24	0.127	0.256	0.531	0.857	1.318	1.711	2.064	2.492	2.797	3.745
25	0.127	0.256	0.531	0.856	1.316	1.708	2.060	2.485	2.787	3.725
26	0.127	0.256	0.531	0.856	1.315	1.706	2.056	2.479	2.779	3.707
27	0.127	0.256	0.531	0.855	1.314	1.703	2.052	2.473	2.771	3.690
28	0.127	0.256	0.530	0.855	1.313	1.701	2.048	2.467	2.763	3.674
29	0.127	0.256	0.530	0.854	1.311	1.699	2.045	2.462	2.756	3.659
30	0.127	0.256	0.530	0.854	1.310	1.697	2.042	2.457	2.750	3.646
40	0.126	0.255	0.529	0.851	1.303	1.684	2.021	2.423	2.704	3.551
60	0.126	0.254	0.527	0.848	1.296	1.671	2.000	2.390	2.660	3.460
120	0.126	0.254	0.526	0.845	1.289	1.658	1.980	2.358	2.617	3.373
∞	0.126	0.253	0.524	0.842	1.282	1.645	1.960	2.326	2.576	3.291

[a] For a given percentage p, the $p\%$ value t_p of the t distribution is defined by the condition

$$P(|t| > t_p) = \frac{p}{100}.$$

Thus the probability that t differs from its mean zero in either direction by more than t_p is equal to $p\%$.

[b] From H. Cramér, "Elements of Probability Theory." Wiley, New York, 1955. Used with permission of the publisher.

Table IV

Critical Values of r for the Sign Test[a,b,c]

N (number of pairs)	1%	5%	10%	25%	N (number of pairs)	1%	5%	10%	25%
1					31	7	9	10	11
2					32	8	9	10	12
3				0	33	8	10	11	12
4				0	34	9	10	11	13
5			0	0	35	9	11	12	13
6		0	0	1	36	9	11	12	14
7		0	0	1	37	10	12	13	14
8	0	0	1	1	38	10	12	13	14
9	0	1	1	2	39	11	12	13	15
10	0	1	1	2	40	11	13	14	15
11	0	1	2	3	41	11	13	14	16
12	1	2	2	3	42	12	14	15	16
13	1	2	3	3	43	12	14	15	17
14	1	2	3	4	44	13	15	16	17
15	2	3	3	4	45	13	15	16	18
16	2	3	4	5	46	13	15	16	18
17	2	4	4	5	47	14	16	17	19
18	3	4	5	6	48	14	16	17	19
19	3	4	5	6	49	15	17	18	19
20	3	5	5	6	50	15	17	18	20
21	4	5	6	7	51	15	18	19	20
22	4	5	6	7	52	16	18	19	21
23	4	6	7	8	53	16	18	20	21
24	5	6	7	8	54	17	19	20	22
25	5	7	7	9	55	17	19	20	22
26	6	7	8	9	56	17	20	21	23
27	6	7	8	10	57	18	20	21	23
28	6	8	9	10	58	18	21	22	24
29	7	8	9	10	59	19	21	22	24
30	7	9	10	11	60	19	21	23	25

Continued

Table IV—Continued

N (number of pairs)	1%	5%	10%	25%	N (number of pairs)	1%	5%	10%	25%
61	20	22	23	25	76	26	28	30	32
62	20	22	24	25	77	26	29	30	32
63	20	23	24	26	78	27	29	31	33
64	21	23	24	26	79	27	30	31	33
65	21	24	25	27	80	28	30	32	34
66	22	24	25	27	81	28	31	32	34
67	22	25	26	28	82	28	31	33	35
68	22	25	26	28	83	29	32	33	35
69	23	25	27	29	84	29	32	33	36
70	23	26	27	29	85	30	32	34	36
71	24	26	28	30	86	30	33	34	37
72	24	27	28	30	87	31	33	35	37
73	25	27	28	31	88	31	34	35	38
74	25	28	29	31	89	31	34	36	38
75	25	28	29	32	90	32	35	36	39

[a] H_0 is accepted if the observed number of the less frequent sign exceeds the value of r in the table.

For values of N larger than 90, approximate values of r may be found by taking the nearest integer less than $(N-1)/2 - k\sqrt{N+1}$, where k is 1.2879, 0.9800, 0.8224, and 0.5752 for the 1, 5, 10, and 25% values, respectively.

[b] Two-tail percentage points for the binomial for $p = 0.5$.

[c] From W. J. Dixon and A. M. Mood, The statistical sign test, *J. Amer. Statist. Assoc.* **41**, 557–566 (1946). Used with permission of the authors and the Editor.

Answers to Selected Problems

Chapter 1

3. $A \cup (A^c \cap B) \cup (A \cup B)^c = D$ (say);
   $A \cup (A^c \cap B) = (A \cup A^c) \cap (A \cup B) = \Omega \cap (A \cup B) = A \cup B$;
   $D = (A \cup B) \cup (A \cup B)^c = \Omega$.
4. $A \cup B = \{5, \ldots, 25\}$, $A \cap B = \{12, \ldots, 20\}$, $A \cap B^c = \{5, \ldots, 1\}$
6. $A^c = \{$same face on both coins$\}$.
7. $(A \cup B) - B = (A \cup B) \cap B^c = (A \cap B^c) \cup (B \cap B^c) = A \cap B^c$.
10. $(A - B) \cap B = (A \cap B^c) \cap B = A \cap (B^c \cap B) = A \cap \varnothing = \varnothing$.
11. $(A \cup B) \cap (A \cup C) = [A \cap (A \cup C)] \cup [B \cap (A \cup C)]$
    $= [A \cup (A \cap C)] \cup [(B \cap A) \cup (B \cap C)]$
    $= A \cup (B \cap A) \cup (B \cap C) = A \cup (B \cap C)$.

ANSWERS TO SELECTED PROBLEMS

14. A. **15.** $A \subset B$.

17. $\Omega = \{\omega_1, \omega_2, \omega_3, \omega_4\}$, where $\omega_1 = (H, H)$, $\omega_2 = (H, T)$, $\omega_3 = (T, H)$, $\omega_4 = (T, T)$. (Here H stands for the outcome "head," T for "tail.") $\mathfrak{U} = \mathfrak{P}(\Omega)$ that is the system of all subsets of Ω. $P(\omega_j) = \frac{1}{6}$ $(j = 1,2,3,4,5,6)$ and for $A \in \mathfrak{U}$, $P(A) = \sum_{\omega_j \in A} P(\omega_j)$.

21. $p_1 = \frac{1}{6}, p_2 = p_3 = p_4 = p_5 = \frac{1}{8}, p_6 = \frac{1}{3}$. **22.** .28.

23. (a) $A \cap B$ is the event that the selected number is 2, its probability is $\frac{1}{10}$.
(b) $A - C = \{2,4,8,10\}$, $P(A - C) = \frac{2}{5}$.

Chapter 2

2. $P(A \cup B \cup C) = P[(A \cup B) \cup C]$
$= P(A \cup B) + P(C) - P[(A \cup B) \cap C]$
$= P(A) + P(B) - P(A \cap B) + P(C)$
$- P[(A \cap C) \cup (B \cap C)]$
$= P(A) + P(B) + P(C) - P(A \cap B) - P(A \cap C)$
$- P(B \cap C) + P(A \cap B \cap C)$.

3. $C = (A \cup B)^c$, $P(C) = 1 - P(A \cup B) = 1 - P(A) - P(B)$, since $P(A \cap B) = 0$.

5. Follows from $A = (A - B) \cup (A \cap B)$.

6. $\Omega = \{(x, y): x, y = b, w, r, y\}$; $\mathfrak{U}$ is the system of all subsets of Ω; $P(x, y) = \frac{1}{16}$.

7. $\frac{1}{10^4}$. **8.** $\frac{1}{7}$. **11.** $\frac{1}{6}$.

13. $A \cap B = \emptyset$ implies $P(A \cap B) = 0$. Independence means $P(A \cap B) = P(A)P(B)$, so that incompatible events A and B are independent if, and only if, either $P(A) = 0$ or $P(B) = 0$.

14. $\frac{1}{4}$. **15.** $1 - P(A^c \cap B^c) = P(A \cup B) = P(A) + P(B) - P(A \cap B)$.

17. (a) $\frac{7}{15}$; (b) $\frac{2}{7}$. **18.** $\frac{1}{5}$.

19. $p = P(H); q = P(T) = 1 - p; p_n = P(\text{terminates at } n\text{th toss}) = pq^{n-1}$.

ANSWERS TO SELECTED PROBLEMS

21. *Hint*: One has $(1 + x)^N = (1 + x)^M (1 + x)^{N-M}$. Apply the binomial theorem to both sides of this equation and compare the coefficients of x^n. In this way you obtain the relation

$$\sum_{k+j=n} \binom{M}{j} \binom{N-M}{k} = \binom{N}{n}$$

and the statement follows then by rewriting the summation on the left-hand side of the last equation.

22. $\dfrac{q}{(1+q)}$. **23.** $\sum\limits_{k=0}^{\infty} q^{2k} p = p \sum\limits_{k=0}^{\infty} (q^2)^k = \dfrac{p}{1-q^2} = \dfrac{1}{1+q}$.

25. $\dfrac{4}{\binom{52}{5}} = \dfrac{1}{5 \cdot 52 \cdot 51 \cdot 49}$. **26.** $\dfrac{47}{85} = .55294$.

27. $p = 1 - \dfrac{365!}{(365-n)! \, 365^n}$, $n = 23$. **28.** $\dfrac{2}{(n-1)}$.

Chapter 3

1. $F_Y(y) = F_X(y - a)$.

2. (i) if $a > 0$, $F_Y(y) = F_X\left(\dfrac{y-b}{a}\right)$, $p_Y(y) = \dfrac{1}{a} p_X\left(\dfrac{y-b}{a}\right)$.

   (ii) if $a < 0$, $F_Y(y) = 1 - F_X\left(\dfrac{y-b}{a}\right)$, $p_Y(y) = -\dfrac{1}{a} p_X\left(\dfrac{y-b}{a}\right)$.

3. $[Y \leq y] = [|X| \leq y] = [-y \leq X \leq y] \in \mathfrak{U}$;
   $F_Y(y) = F_X(y) - F_X(-y - 0)$.

4. $F_Y(y) = F_X(y^2)$, $p_Y(y) = \begin{cases} 2y p_X(y^2) & \text{for } y > 0, \\ 0 & \text{otherwise.} \end{cases}$

5. $F_Y(y) = \begin{cases} 0 & \text{for } y \leq 0, \\ 1 - e^{-\theta\sqrt{y}} & \text{for } y > 0, \end{cases}$ $p_Y(y) = \dfrac{\theta}{2} y^{-1/2} e^{-\theta\sqrt{y}}$ for $y > 0$.

6. $F_{I_A}(x) = \begin{cases} 0 & \text{if } x < 0, \\ 1 - P(A) & \text{if } 0 \leq x < 1, \\ 1 & \text{if } x \geq 1. \end{cases}$

7. $k = \dfrac{(1+a)(1+b)}{b-a}$.

9. $p_X(x) = p_Y(x) = \begin{cases} 2x & \text{if } 0 < x \le 1, \\ 0 & \text{otherwise,} \end{cases}$

since

$$\int_0^1 \int_0^1 p(x,y)\,dx\,dy = 4\int_0^1 x\,dx \int_0^1 y\,dy = 1.$$

$p_X(x) = 2x \quad (0 < x < 1); \qquad p_Y(y) = 2y \quad (0 < y < 1).$

$F_X(x) = \begin{cases} x^2 & \text{if } 0 < x < 1, \\ 0 & \text{otherwise;} \end{cases} \qquad F_Y(y) = \begin{cases} y^2 & \text{if } 0 < y < 1, \\ 0 & \text{otherwise.} \end{cases}$

10. $k = \tfrac{3}{2}$.

11. $p_X(x) = \begin{cases} xe^{-x} & \text{for } x > 0, \\ 0 & \text{otherwise;} \end{cases}$

$p_Y(y) = \begin{cases} e^{-y} & \text{for } y > 0, \\ 0 & \text{otherwise,} \end{cases}$

hence $p_{XY}(x,y) = p_X(x)\,p_Y(y)$ so that X and Y are independent.

12. X and Y are not independent, since

$$p_X(x) = \begin{cases} 2(1-x) & \text{for } 0 < x \le 1, \\ 0 & \text{otherwise;} \end{cases}$$

$$p_Y(y) = \begin{cases} 2y & \text{for } 0 < y < 1, \\ 0 & \text{otherwise.} \end{cases}$$

13. $1 - \dfrac{1}{\sqrt{2}}$. 14. $\tfrac{1}{2}$. 15. 0.13.

17. (a) $\tfrac{1}{10}$; (b) $\tfrac{7}{75}$.

18. $p_Y(y) = \begin{cases} 0 & \text{if } y \le 1, \\ \dfrac{1}{y^2} & \text{if } y > 1. \end{cases}$

19. (a) $\dfrac{39m - m^2}{380}$; (b) $m = 6$. 20. (a) $1 - (\tfrac{9}{10})^m$; (b) $m = 7$.

21. $\dfrac{1}{\pi(1+x^2)}$.

ANSWERS TO SELECTED PROBLEMS

Chapter 4

1. 7. **2.** $\mathscr{E}(X) = \dfrac{1}{\ln 2}$. **3.** np.

4. λ. **5.** $\dfrac{b+a}{2}$. **6.** 0.

7. $\dfrac{1}{\theta}$. **8.** $\dfrac{2}{p}$. **9.** $\dfrac{k}{p}$.

10. $\dfrac{Mn}{N}$. **14.** $P(A)[1 - P(A)]$. **15.** $\sigma\sqrt{\dfrac{2}{\pi}}$.

16. $\alpha_k = \dfrac{m}{m-k}$ exists for $k < m$.

17. $\alpha_{2k} = \dfrac{r^{2k}}{2k+1}$, $\alpha_{2k-1} = 0$, $\beta_k = \dfrac{r^k}{k+1}$.

18. $\text{Var}(X+Y) = \text{Var}(X) + \text{Var}(Y) + 2\,\text{Cov}(X, Y)$.

19. Not independent, but uncorrelated.

20. Let $A = [g(x) \geq M]$ and suppose that $F(x)$ is absolutely continuous with density $p(x)$. Then $\mathscr{E}[g(X)] \geq \int_A g(x)p(x)\,dx \geq MP(A) = MP[g(X) \geq M]$.

22. $\frac{1}{2}(1+x)$. **23.** From normal table .0456; from Chebyshev inequality .25.

24. (a) $N(ps - c)$; (b) 63; (c) first process preferable.

25. From Chebyshev's inequality approximately .2613; from normal table .05.

26. $\Phi(-3) = 1 - \Phi(3) = .0013$.

27. $\mathscr{E}(e^{X^2}) = \displaystyle\int_{-\infty}^{\infty} (\exp x^2) p(x)\,dx \geq \int_{x^2 \geq \varepsilon^2} (\exp x^2) p(x)\,dx$

$\geq (\exp \varepsilon^2) \displaystyle\int_{x^2 \geq \varepsilon^2} p(x)\,dx$

$= (\exp \varepsilon^2)\, P(X^2 \geq \varepsilon^2) = (\exp \varepsilon^2)\, P(|X| \geq \varepsilon)$.

29. X has Poisson distribution with $\mathscr{E}(X) = 6$. $P(0 \le X < 9) = 1 - P(X \ge 9)$ and apply the inequality of Problem 20.

Chapter 5

1. Yes, by Theorem 5.1.2.   **2.** Yes, by Theorem 5.1.2.
3. By Chebyshev's inequality $n \ge 250000$; by central limit theorem $n \sim 16641$.
4. Theoretical frequencies 108.7, 66.3, 20.2, 4.1, 0.7.
5. Use Theorem 5.1.2.   **6.** $s < \frac{1}{2}$.   **7.** $s < \frac{1}{2}$.
8. Sequence obeys the law of large numbers.
9. The central limit theorem holds.
10. The central limit theorem holds.
11. 0.975.
12. (a) 0.2503; (b) 0.2648; (c) 0.164.
13. (a) 0.0013; (b) 0.0009; (c) 0.0010.

Chapter 6

1. $Z = X + Y$; $p_Z(z) = \begin{cases} z & \text{if } 0 < z < 1, \\ 2 - z & \text{if } 1 \le z < 2, \\ 0 & \text{otherwise.} \end{cases}$

2. $Z = X + Y$; $p_Z(z) = \begin{cases} 0 & \text{if } z < 0, \\ \dfrac{1}{2^{(m+n)/2}\, \Gamma\!\left(\dfrac{m+n}{2}\right)} z^{[(n+m)/2]-1} e^{-z/2}. \end{cases}$

4. $p_Y(y) = \dfrac{2\Gamma\!\left(\dfrac{n+1}{2}\right)}{\Gamma\!\left(\dfrac{n}{2}\right)\sqrt{\pi n}} \left(1 + \dfrac{y^2}{n}\right)^{-(n+1)/2}.$

5. (a) 2.764; (b) 3.169.   **6.** 24.996.

ANSWERS TO SELECTED PROBLEMS   233

7. $p_Y(y) = \begin{cases} 0 & \text{if } y < 0, \\ \dfrac{\Gamma\left(\dfrac{n+1}{2}\right)}{\Gamma\left(\dfrac{n}{2}\right)\sqrt{n\pi}} \left(1 + \dfrac{y}{n}\right)^{-(n+1)/2} y^{-1/2} & \text{if } y > 0. \end{cases}$

8. Special case of problem 7 for $n = 1$.

9. $\begin{cases} \dfrac{1}{2^{n/2}\,\Gamma\left(\dfrac{n}{2}\right)} \exp\left[-\dfrac{1}{2}\dfrac{y}{1-y}\right] y^{(n/2)-1}(1-y)^{-(n/2)-1} & \text{if } 0 < y < 1, \\ 0 & \text{otherwise.} \end{cases}$

11. Poisson with parameter 2λ.   **12.** Poisson with parameter $\lambda + \mu$.

13. Poisson with parameter $n\lambda$.   **14.** $p(x) = \theta^2 x e^{-\theta x}$.

17. $\alpha_1 = 0$, $\beta_1 = \sqrt{2}$, second moment does not exist.

18. $\text{Var}(X) = 3$.   **19.** $n(n+2)(n+4)$.

20. $q(x) = \begin{cases} 0 & \text{if } x < -1, \\ \tfrac{1}{2}(x+1)^2 & \text{if } -1 < x < 0, \\ \tfrac{1}{2} + x - x^2 & \text{if } 0 < x < 1, \\ \tfrac{1}{2}(x-2)^2 & \text{if } 1 < x < 2, \\ 0 & \text{if } x > 2. \end{cases}$

Chapter 7

1. $\bar{X}_n = \dfrac{1}{n} S_n$, $P\left(\bar{X}_n = \dfrac{k}{n}\right) = \dfrac{e^{-\lambda n}(\lambda n)^k}{k!}$.

2. $\mathcal{E}(\bar{X}_n) = \lambda$, $\text{Var}(\bar{X}_n) = \dfrac{\lambda}{n}$, $\mathcal{E}(\bar{X}_n^2) = \lambda^2 + \dfrac{\lambda}{n}$.

3. $\dfrac{n_1 \bar{X}_1 + n_2 \bar{X}_2}{n_1 + n_2}$.

4. $s^2 = \dfrac{n_1 s_1^2 + n_2 s_2^2}{n_1 + n_2} + \dfrac{n_1 n_2}{(n_1 + n_2)^2}(\bar{X}_1 - \bar{X}_2)^2$.

5. $\bar{X} = 0.6825$; $s^2 = 0.8117$; $a_3 = 1.0876$.

6. $\mathcal{E}(\bar{X}) = 0$; $\text{Var}(\bar{X}) = \dfrac{1}{3n}$.

7. $\mathcal{E}(\bar{X}) = \tfrac{1}{2}(b + a)$; $\text{Var}(\bar{X}) = \dfrac{(b-a)^2}{12n}$.

8. $\mathcal{E}(\bar{X}) = 0$; $\text{Var}(\bar{X}) = \dfrac{1}{2n}$.

9. Method A: $\bar{X} = 80.02$, $s^2 = 0.00052$; Method B: $\bar{X} = 79.98$, $s^2 = 0.00097$.

11. $\mathcal{E}(\bar{X}) = n$; $\text{Var}(\bar{X}) = \dfrac{2n}{m}$. 12. $\mathcal{E}(\bar{X}) = \theta^{-1}$, $\text{Var}(\bar{X}) = (n\theta^2)^{-1}$.

13. Does not exist. 14. $\mathcal{E}(\bar{X}) = \dfrac{\lambda}{\theta}$: $\text{Var}(\bar{X}) = \dfrac{\lambda}{n\theta^2}$.

Chapter 8

1. (a) $\mathcal{E}(\theta_n) = \sum\limits_{k=1}^{n} a_{nk}\mathcal{E}(X_k) = \mathcal{E}(X_k)$; (b) $P(|\theta_n - \theta| \leq \varepsilon)$
$\geq 1 - \dfrac{\text{Var}(\theta_n)}{\varepsilon^2}$; $\text{Var}(\theta_n) = \sigma^2 \sum\limits_{k=1}^{n} a_{nk}^2 \to 0$ as $n \to \infty$,
hence $\lim\limits_{n \to \infty} P(|\theta_n - \theta| \leq \varepsilon) = 1$.

3. $\mathcal{E}(\bar{X}) = \lambda$; $\mathcal{E}\left(\dfrac{ns^2}{n-1}\right) = \lambda$. 4. $\tilde{\theta} = \dfrac{\bar{X}}{s^2}$; $\tilde{\lambda} = \dfrac{\bar{X}^2}{s^2}$,

5. $\hat{\theta} = \dfrac{\lambda}{\bar{X}}$. 6. $\hat{\theta} = \dfrac{1}{\bar{X}}$. 7. $\tilde{\theta} = \dfrac{1}{\bar{X}}$.

8. $\tilde{\theta} = \dfrac{1}{s}$; $\tilde{\mu} = \bar{X} - s$. 9. $\tilde{r} = \sqrt{2s^2}$.

10. (a) $\hat{k} = 3\bar{X}$; (b) $\sum\limits_{j=1}^{n}(k - |x_j|)^{-1} - 2nk^{-1} = 0$.

11. $\hat{X} = 540$. 12. $\bar{X} = .1257$; $(.1206, .1308)$.

14. $(.017, .063)$. 16. $(2.520, 5.699)$.

Chapter 9

1. Mean of Treatment (1): $+0.75$, of Treatment (2): $+2.33$, of $(2)-(1)$ $+1.58$. Standard deviation of Treatment (1): 1.70, of Treatment (2): 1.90, of $(2)-(1)$: 1.17; H_0: no difference between treatments [that is, mean of the difference $(2)-(1)$ is zero], $t = 1.99$ H_0 rejected at 5%. By sign test: 9 positive differences 1 tie, H_0 rejected.
3. $H_0 : \theta_0 = 0$, $t = 1.32$ no effect of Treatment 1; $t = 3.67$, Treatment 2 effective.
4. $t = 3.30$, 19 degrees of freedom. 5% value of $t = 2.09$ (1% value $= 2.861$). Hypothesis that there is no difference is rejected.
5. The t-test (or the sign test) does not reject the hypothesis that batteries of the two manufacturers do not differ in capacity.
6. $\chi^2 = 4.62$, 3 degrees of freedom. The null hypothesis that the probability of failure is the same for all four types of meters is not rejected.
7. $\bar{X} = -8.22$; $s = 9.746$; $t = -3.47$. The hypothesis that there are no changes is rejected.
9. Difference significant at 5% level.
11. The chi-square test rejects the hypothesis at the 1% level.
13. (a) $P\left(\bar{X} > \dfrac{1.645}{\sqrt{n}}\right) = .05$ by Table I or IA respectively;
    (b) $|\bar{X}| > \dfrac{2.326}{\sqrt{n}}$; $\bar{X} > \dfrac{2.054}{\sqrt{n}}$.
14. $\pi(\theta) = 1 - \Phi(1.645 - 2\theta)$.
15. $\chi^2 = .72$ at 1 degree of freedom. The hypothesis that coin is unbiased is not rejected at 5% or 1% level.
16. $\chi^2 = 19.7$ at 5 degrees of freedom. The hypothesis that die is not loaded is rejected at 5% or 1% level.

Index

A

Absolutely continuous distribution
   function, 43
   example of, 48–54
Absolutely continuous random variable, 44
Accept hypothesis, 176, 179
Addition
   of independent normal random variables, 121
   of independent random variables, 121, 209
Addition rule, 21
   for expectations, 81

Additive, countably, 20
Alternative hypothesis, 179
Alternatives (simple), 31
Asymptotic confidence interval, 169, 170
Asymptotically efficient, 159
Asymptotically normal, 146
Axioms, 2
   of probability theory, 14, 15

B

Bayes theorem, 28
Bernoulli trial, 32, 48, 108, 116
Bernoulli's law of large numbers, 105
Binomial coefficient, 201

Binomial distribution, 47, 187
Binomial probabilities, 32, 108, 111
Binomial theorem, 202
Bivariate distribution, 56
Bivariate normal distribution, 63
Boole's inequality, 22
Bounded random variable, 75

C

C_3, 207
Cardano, G., 4
Cauchy distribution, 54, 74, 132, 145, 146
Central limit theorem, 108, 113, 146, 207, 213
  for Bernoulli trials, 112
Central sample moments, 140
Certain event, 9
Chebyshev's inequality, 95, 102
Chi-square distribution, 123–129, 188
  as special case of gamma distribution, 129, 147
Coefficient of correlation, 92, 93, 95
Combinations, 201
Combinatorial formulas, 199
Complementary event, 8
Completely additive, 15
Completely independent, 30
Composite hypothesis, 176
Compound event, 8
Conditional expectation, 97, 98
Conditional frequency, 59
  of bivariate normal, 63, 65
Conditional frequency function, 59, 96, 97
Conditional probability, 24–27, 57
  discrete, 57
Confidence coefficient, 165
Confidence interval, 163, 164
Consistency, 155, 159
Consistent sequence of estimates, 155, 159
Continuity point, 43
Contraction operator, 208
Convolution, 209

Correlation, coefficient of, 92, 93, 95
Countable, 12
Countably additive, 20
Covariance, 91, 95
Cramér, H., 69, 117, 151, 173, 197, 222, 223
Critical region, 177
Czuber, E., 4

D

Decreasing sequence of events, 23
Degenerate distribution, 45
Degenerate random variable, 45
Degrees of freedom, 124, 130
  of χ^2 distribution, 124
  of Student's distribution, 130
DeMoivre, 108
Density function, 44
Denumerable, 12
Difference of random variables, 40
Discontinuity point, 43
Discrete distribution functions, 43
  examples of, 44–48
Discrete random variable, 44
Dispersion, 85
Distribution
  binomial, 47, 187
  bivariate normal, 63
  Cauchy, 54, 74, 132, 145, 146
  chi-square, 123–129, 188
  degenerate, 45
  exponential, 51, 129
  gamma, 129
  Gauss–Laplace, 51
  Gaussian, 51
  geometric, 48, 72, 87
  hypergeometric, 47
  Laplace, 51
  Moivre–Laplace, 51
  multinomial, 61, 189
  multivariate discrete, 57, 60, 61
  multivariate normal, 62–66
  negative exponential, 51
  normal, 51, 88, 74, 87, 94
  Pascal, 48

INDEX

Poisson, 48, 114
rectangular, 49
of sample characteristics, 144–148
Simpson's, 50
Student's, 129–132
of sum of independent absolutely continuous random variables, 119
t-distribution, 129–132
triangular, 50
uniform, 49
Distribution free methods, 186
Distribution function, 40–54
absolutely continuous, examples of, 48–54
discrete, examples of, 44–48
multivariate, 54–66
univariate, 40–54
of sample characteristics, 144–148
Dixon, W. J., 225
Drawn at random, 18, 31

E

Efficient estimate, 157, 159
Estimates, properties of, 154–157
Estimation, 153–169
by interval, 162–169
Exhaustive set of events, 11
Expectation, 71
of functions of random variables, 76–81
for multivariate distribution, 74, 75
Expected value, 72
Exponential distribution, 51, 129
Event
complementary, 8
compound, 8
opposite, 8
simple, 7

F

Factorial, 199
Feller, W., 218
Fermat, P., 4

Finite additivity, 20
Finite probability spaces, 31–33
Finite second moment, 90
Fisher, R. A., 157
Frequency function, 44
conditional, 59, 63, 65, 96, 97
Function of random variables, 76
Fundamental concepts, 2

G

Galilei, G., 4
Gamma distribution, 129
Gamma function, 205–206
Gauss–Laplace distribution, 51
Gaussian distribution, 51
Geometric distribution, 48, 72, 87
Gosset, W. S., 129

H

Halmos, P. R., 18
Hilbert, D., 1, 4
Hypergeometric distribution, 47
Hypergeometric probabilities, 34
Hypothesis
composite, 176
simple, 176

I

Identically distributed random variables, 113
Implication of events, 10
Implication rule, 22
Impossible event, 9
Incompatible events, 10
Increasing sequence of events, 23
Independent events, 29, 30
Independent random variables, 66
Indicator random variables, 45, 120
Intersection, 8
Interval estimation, 162–169

J

Joint distribution, 55
Jump, 42

K

Kamke, E., 18
Kolmogorov, A. N., 1, 4

L

Laplace distribution, 51
Law of large numbers, 105–107
Likelihood equation, 158
Likelihood function, 158
Limit theorems, 105–114, 213–218
Lindeberg condition, 214, 218
Location, 85

M

$\mathscr{M}$ (family of functions), 77
$\mathscr{M}_n$, 80
Marginal (frequencies), 58, 59
Marginal probability, 57
Mathematical expectation, 71–74
Mathematical statistics, 135
Maximum likelihood estimate, 158
Maximum likelihood method, 158
Mean, 53, 72
Measure
  of location, 85
  of scatter, 85
Median, 4, 18
Method of moments, 161, 162
Mixed moment, 90
Moivre–Laplace distribution, 51
Moments
  absolute, 84
  algebraic, 84
  central, 84
  of sample characteristics, 139, 140

Mood, A. M., 220, 221, 225
Multinomial distribution, 61, 189
Multinomial probabilities, 61
Multiplication rule, 27
  for expectations, 82–83
Multivariate discrete distributions, 57, 60, 61
Multivariate distribution, 54–67
Multivariate normal distribution, 62–66
Mutually exclusive events, 10

N

n-dimensional distribution, 56
Natrella, M., 152, 173, 198
Negative exponential distribution, 51
Neyman, J., 184
Nonnegative random variables, 83
Nonparametric methods, 186–191
Norm (of function), 208
Normal distribution, 51, 74, 87, 88, 94
Null hypothesis, 177

O

Operators, 207–209
Opposite event, 9, 10
Ore, O., 4
Outcome Space, 7

P

Pairwise independent, 30
Paramater space, 176
Pascal, B., 4
Pascal distribution, 48
Pearson, Egon, 184
Pearson, Karl, 161, 188, 189
Permutations, 199
Point estimation, 157–162
Point of increase, 43
Poisson approximation to binomial, 113–114

INDEX

Poisson distribution, 48, 114
Poisson trials, 107
Population, 153
Population distribution function, 153
Population regression coefficient, 98, 141
Positive definite quadratic form, 62
Possible value, 43
Power function, 182–183
Power test, 178–182
Probabilities, 11–14
    conditional, 24–27, 57
Probability distribution, 37
Probability operator, 210–213
Probability space, 15
    denumerable, 16
    finite, 16, 31
Product moment, 90
Product of random variables, 40

R

Random, 18
Random phenomenon, 3
Random variables, 37–40
Realization of sample, 138
Reciprocal of random variables, 40
Rectangular distribution, 49
Region of rejection, 177
Regression, 96–100
Regression coefficient, 97, 98, 141
Regression curve, 97
Regression line, 98, 141
Reject hypothesis, 176, 179
Relative efficiency, 156
Relative frequency, 3, 11
Right continuous, 41

S

Saltus, 42
Sample, 138
    characteristics, 139
    distribution functions, 139
    mean, 139
    moments, 139, 140
    regression, 141
    regression coefficient, 141
    regression line, 142
    size, 138, 177
    space, 177
    variance, 140
Sampling
    with replacement, 33
    without replacement, 33
Scatter, 85
Schwarz inequality, 89
Second law of Laplace, 51
Selection at random, 18
Set theoretic operations, 9
σ-algebra, 15
σ-field, 15
Sign test, 187–188
Significance level, 177
Simple hypothesis, 176
Simpson's distribution, 50
Size of critical region, 177
Standardized random variable, 88
Standardized normal distribution, 52
Statistic, 138
Statistical data, 137
Statistical hypothesis, 176
Statistical regularity, 3
Stirling's formula, 109, 203
Stochastic independence, 147
Stoll, R. R. 18
"Student," 152, 198
Student's distribution, 129–132
Sum, of random variables, 40
Square of random variables, 40

T

t-distribution, 129–132
t-test, 184–186
Taylor's theorem, 108
Testing hypothesis, 175–191
Tests of goodness-of-fit, 188
Todhunter, I., 4
Total probability rule, 27

Triangular distribution, 50
Trotter, H. F., 218
True (coin, die), 17
Tucker, H. G., 69, 103, 117, 152
Type I error, 179
Type II error, 179
Typical value, 71, 139

U

Unbiased, 17
  coin, die, 17, 34
Uncorrelated random variable, 92
Uniform distribution, 49

Uniform probabilities, 16, 31
Uniformly most powerful region, 183
Union, 8
Univariate distribution, 56

V

Variance, 53, 86

W

Weyl, H., 4
Widder, D. V., 205

ST. MARY'S COLLEGE OF MARYLAND LIBRARY
ST. MARY'S CITY, MARYLAND

C65076

DATE DUE			